Universitext

Universitext

Universitext is a series of textbooks that presents material from a wide variety of mathematical disciplines at master's level and beyond. The books, often well class-tested by their author, may have an informal, personal, even experimental approach to their subject matter. Some of the most successful and established books in the series have evolved through several editions, always following the evolution of teaching curricula, into very polished texts.

Thus as research topics trickle down into graduate-level teaching, first textbooks written for new, cutting-edge courses may make their way into *Universitext*.

For further volumes:
http://www.springer.com/series/223

Mark de Longueville

A Course in Topological Combinatorics

 Springer

Mark de Longueville
Hochschule für Technik und Wirtschaft Berlin
University of Applied Sciences
Berlin, Germany

ISSN 0172-5939   ISSN 2191-6675 (electronic)
ISBN 978-1-4899-8826-3   ISBN 978-1-4419-7910-0 (eBook)
DOI 10.1007/978-1-4419-7910-0
Springer New York Heidelberg Dordrecht London

Mathematics Subject Classification (2010): 05-01, 52-01, 55M99, 55M35, 05C15, 05C10, 91B08, 91B32

Printed on acid-free paper

Springer is part of Springer Science+Business Media (www.springer.com)

For L., with whom I fell, first from the bicycle and then in love

Preface

Topological combinatorics is a very young and exciting field of research in mathematics. It is mostly concerned with the application of the many powerful tools of algebraic topology to combinatorial problems. One of its early landmarks was Lovász's proof of the Kneser conjecture published in 1978. The combination of the two mathematical fields—topology and combinatorics—has led to many surprising and elegant proofs and results.

In this textbook I present some of the most beautiful and accessible results from topological combinatorics. It grew out of several courses that I have taught at Freie Universität Berlin, and is based on my personal taste and what I believe is suitable for the classroom. In particular, it aims for a clear and vivid presentation rather than encyclopedic completeness.

The text is designed for an advanced undergraduate level. Primarily it serves as a basis for a course, but is written in such a way that it just as well may be read by students independently. The textbook is essentially self-contained. Only some basic mathematical experience and knowledge—in particular some linear algebra—is required. An extensive appendix allows the instructor to design courses for students with very different prerequisites. Some of those designs will be sketched later on.

The textbook has four main chapters and several appendices. Each chapter ends with an accompanying and complementing set of exercises. The main chapters are mostly independent of each other and thus allow considerable flexibility for an individual course design. The dependencies are roughly as follows.

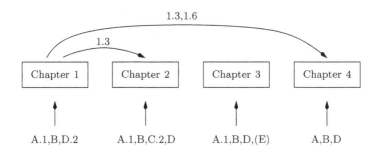

Suggested Course Outlines

For students with previous knowledge of graph theory and the basics of algebraic topology including simplicial homology theory. Use Chaps. 1–4. Whenever concepts and results on partially ordered sets and their topology from Appendix C or on group actions from Appendix D are missing, they should be included. Oliver's Theorem 3.17, which is proven in Appendix E, can easily be used as a black box. If the students are experienced with homology and if time permits, I recommend studying Appendix E after Chap. 3.

For students with previous knowledge of the basics of algebraic topology including simplicial homology theory only. Proceed as in the last case and provide the basics of graph theory from Appendix A along the way.

For students with previous knowledge of graph theory only. I recommend that the instructor introduces some basic topology with Sects. B.1 and B.3, and then presents Chap. 1, skipping the homological proofs. Before Sect. 1.6 I recommend giving a topology crash course with Sects. B.4–B.9. Proceed with Chaps. 2–4 and add concepts and results from Appendices C and D as needed. Apply Theorem 3.17 as a black box and use Appendix E as a motivation to convince students to study algebraic topology.

For motivated students with neither graph theory nor algebraic topology knowledge. Proceed as in the last case and provide the basics from graph theory from Appendix A along the way.

Acknowledgments

First of all, I would like to thank all the authors of research papers and textbooks—several of them I know personally—on which this book is based. I am thankful to Martin Aigner for helpful advice and for supporting the initial idea of the project, and to Günter M. Ziegler for providing excellent working conditions in his research group at Technische Universität Berlin. I am indebted to all the students who took part in my courses on the subject, and to all of the colleagues who helped me with discussions, suggestions, and proofreading. In particular, I want to thank Anna Gundert, Nicolina Hauke, Daria Schymura, Felix Breuer, Aaron Dall, Anton Dochtermann, Frederik von Heymann, Frank Lutz, Benjamin Matschke, Marc Pfetsch, and Carsten Schultz. I also want to thank David Kramer from Springer New York for his very helpful copy editing, and finally, Hans Koelsch and Kaitlin Leach from Springer New York for their competent support of this project.

Berlin, Germany   Mark de Longueville

Contents

List of Symbols and Typical Notation

$[n] = \{1, 2, \ldots, n\}$ the set of natural numbers from 1 to n

$|S|$ the number of elements of a set S

$\lfloor x \rfloor$ the largest integer less than or equal to x

$k \mid n$ notation for "k divides n"

$\subseteq$ the subset relation

$\subset$ the proper subset relation

$S = S_1 \dot{\cup} \cdots \dot{\cup} S_n$ a partition of the set S, i.e., $S = S_1 \cup \cdots \cup S_n$ and $S_i \cap S_j = \emptyset$ for all $i \neq j$

$\binom{n}{k} = \frac{n!}{k!(n-k)!}$ the number of k-element subsets of an n-element set

$\binom{X}{k} = \{S \subseteq X : |S| = k\}$ the set of k-element subsets of a set X

$X + Y = X \times \{0\} \cup Y \times \{1\}$ the sum of sets X and Y

K, L abstract simplicial complexes

Δ, Γ geometric simplicial complexes

$\tau \leq \sigma$ notation for "the simplex τ is a face of the simplex σ"

$\sigma^n = \operatorname{conv}(\{e_1, \ldots, e_{n+1}\})$ the standard geometric n-simplex

Δ^n the geometric simplicial complex given by σ^n and all its faces

$K(\Delta)$ the abstract simplicial complex associated with the geometric complex Δ (cf. page 176)

$|\Delta|$ the polyhedron of the geometric simplicial complex Δ

$|K|$ a geometric realization of the abstract complex K or its polyhedron

$\mathcal{P}(X)$ the power set of X, i.e., $\mathcal{P}(X) = \{A : A \subseteq X\}$

2^X will be identified with the power set of X

$2^{[n]}$ will be identified with the power set of $[n]$

2^σ will be identified with the power set of σ, in this notation refers to the abstract simplicial complex given by the simplex σ and all its faces

$\|\cdot\| = \|\cdot\|_2$ the Euclidean norm

$\|\cdot\|_\omega$ the maximum norm

$\mathbb{B}^n = \{x \in \mathbb{R}^n : \|x\| \leq 1\}$ the n-dimensional unit ball

$\mathbb{S}^{n-1} = \{x \in \mathbb{R}^n : \|x\| = 1\}$ the $(n-1)$-dimensional unit sphere

$Q^{n+1} = \operatorname{conv}(\{\pm e_1, \ldots, \pm e_{n+1}\})$ the $(n+1)$-dimensional cross polytope

Γ^n the n-dimensional geometric simplicial complex associated with the boundary of the cross polytope Q^{n+1}

Chapter 1
Fair-Division Problems

Almost every day, we encounter fair-division problems: in the guise of dividing a piece of cake, slicing a ham sandwich, or by dividing our time with respect to the needs and expectations of family, friends, work, etc.

The mathematics of such fair-division problems will serve us as a first representative example for the interplay between combinatorics and topology.

In this chapter we will consider two important concepts: *envy-free fair division* and *consensus division*. These concepts lead to different topological tools that we may apply. On the one hand, there is Brouwer's fixed-point theorem, and on the other hand, there is the theorem of Borsuk and Ulam. These topological results surprisingly turn out to have combinatorial analogues: the lemmas of Sperner and Tucker. Very similar in nature, they guarantee a simplex with a certain labeling in a labeled simplicial complex.

The chapter is organized in such a way that we will discuss in turn a topological result, its combinatorial analogue, and the corresponding fair-division problem.

1.1 Brouwer's Fixed-Point Theorem and Sperner's Lemma

Brouwer's fixed-point theorem states that any continuous map from a ball of any dimension to itself has a fixed point. In two dimensions this can be illustrated as follows. Take two identical maps of Berlin or any other ball-shaped city. Now crumple one of the maps as you like and throw it on the other, flat, map as shown in Fig. 1.1. Then there exists a location in the city that on the crumpled map is exactly above the same place on the flat map.

For the general formulation of Brouwer's theorem, recall that the n-dimensional Euclidean ball is given by all points of distance at most 1 from the origin in n-dimensional Euclidean space, i.e.,

$$\mathbb{B}^n = \{x \in \mathbb{R}^n : \|x\| \le 1\}.$$

M. de Longueville, *A Course in Topological Combinatorics*, Universitext,
DOI 10.1007/978-1-4419-7910-0_1,
© Springer Science+Business Media New York 2013

Fig. 1.1 A city map twice

Theorem 1.1 (Brouwer). *Every continuous map $f : \mathbb{B}^n \to \mathbb{B}^n$ from the n-dimensional ball $\mathbb{B}^n$ to itself has a fixed point, i.e., there exists an $x \in \mathbb{B}^n$ such that $f(x) = x$.*

The first proof that we provide for this theorem relies on a beautiful combinatorial lemma that we will discuss in the next section. There also exists a very short and simple proof using homology theory that we present on page 6.

Sperner's Lemma

Brouwer's fixed-point theorem is intimately related to a combinatorial lemma by Sperner that deals with labelings of triangulations of the simplex. Consider the standard n-simplex given as the convex hull of the standard basis vectors $\{e_1, \ldots, e_{n+1}\} \subseteq \mathbb{R}^{n+1}$, see Fig. 1.2:

$$\sigma^n = \text{conv}\left(\{e_1, \ldots, e_{n+1}\}\right)$$

$$= \left\{ t_1 e_1 + \cdots + t_{n+1} e_{n+1} : t_i \geq 0, \sum_{i=1}^{n+1} t_i = 1 \right\}$$

$$= \left\{ (t_1, \ldots, t_{n+1}) : t_i \geq 0, \sum_{i=1}^{n+1} t_i = 1 \right\}.$$

By Δ^n we denote the (geometric) simplicial complex given by σ^n and all its faces, i.e., $\Delta^n = \{\tau : \tau \leq \sigma^n\}$. Assume that K is a subdivision of Δ^n. We may think of K as being obtained from Δ^n by adding extra vertices. For precise definitions and more details on simplicial complexes we refer to Appendix B. For any n, denote the set $\{1, \ldots, n\}$ by $[n]$. In the definition of a Sperner labeling we will use labels from 1 to $n + 1$, i.e., labels from the set $[n + 1]$.

Definition 1.2. A *Sperner labeling* is a labeling $\lambda : \text{vert}(K) \to [n + 1]$ of the vertices of K satisfying

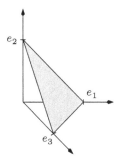

Fig. 1.2 The standard 2-simplex $\sigma^2 = \mathrm{conv}(\{e_1, e_2, e_3\})$

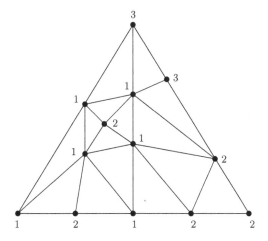

Fig. 1.3 A Sperner-labeled triangulation of a 2-simplex

$$\lambda(v) \in \{i \in [n+1] : v_i \neq 0\}$$

for all $v \in \mathrm{vert}(K)$.

More intuitively, a Sperner labeling is the following. Consider the minimal face of Δ^n that contains v. Say it is given by the convex hull of $e_{i_1}, \ldots, e_{i_k}$. Then v is allowed to get labels only from $\{i_1, \ldots, i_k\}$. In particular, the vertices e_i obtain label i, while a vertex along the edge spanned by e_i and e_j obtains the label i or j, and so on. For an illustration see Fig. 1.3.

Call an n-simplex of K *fully labeled* (with respect to λ) if its $n+1$ vertices obtain distinct labels, i.e., if all possible labels from the set $[n+1]$ appear.

Lemma 1.3 (Sperner [Spe28]). *Let $\lambda : \mathrm{vert}(K) \to [n+1]$ be a Sperner labeling of a triangulation K of the n-dimensional simplex. Then there exists a fully labeled n-simplex in K. More precisely, the number of fully labeled n-simplices is odd.*

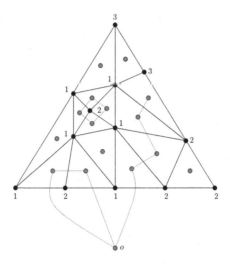

Fig. 1.4 The graph associated to a Sperner labeling

We will now present two inductive proofs of this amazing lemma. The first is a combinatorial construction that constructs one of the desired simplices, while the other is algebraic and uses the concept of a chain complex of a simplicial complex. The inductive proofs reveal a typical phenomenon: while we are mainly interested in the existence of a fully labeled simplex, the induction works only for the stronger statement that there is an odd number of fully labeled simplices.

A third proof, given on page 7 of this section, proves the Sperner lemma as an application of Brouwer's fixed-point theorem.

Proof (combinatorial). The lemma is clearly valid for $n = 1$. Now let $n \geq 2$ and consider the $(n-1)$-dimensional face τ of Δ^n given by the convex hull of $e_1, \ldots, e_n$. Note that K restricted to τ is Sperner labeled with label set $[n]$. We construct a graph as follows. Let the vertex set be all n-simplices of K plus one extra vertex o. The extra vertex o is connected by an edge to all n-simplices that have an $(n-1)$-simplex as a face that is labeled with all labels of $[n]$ and lies within τ. Two n-simplices are connected by an edge if and only if they share an $(n-1)$-dimensional face labeled with all of $[n]$. See Fig. 1.4 for an example of the resulting graph.

By the induction hypothesis, the vertex o has odd degree, since there is an odd number of fully labeled simplices in the labeling restricted to τ. All the other vertices have degree zero, one, or two. To see this, consider the set of labels an n-simplex obtains: either it does not contain $[n]$, it is $[n + 1]$, or it is $[n]$. In the first case, the simplex has degree zero; in the second, it has degree one; and in the last case, it has degree two, since exactly two faces obtain all of $[n]$ as label set; compare Fig. 1.5.

Hence the vertices of degree one other than o (which may have degree one) correspond to the fully labeled simplices. Now, the number of vertices of odd degree

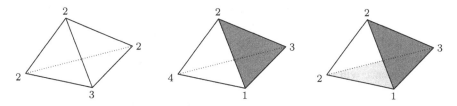

Fig. 1.5 An example of labeled n-simplices of degrees 0, 1, and 2 in the case $n = 3$

in a graph is even. (This is easy to prove; cf. Corollary A.2 in Appendix A.) Since the degree of o is odd, there remains an odd number of fully labeled simplices. □

Proof (algebraic). We proceed by induction. The case $n = 1$ is an easy exercise. Now assume $n \geq 2$. The labeling λ induces a simplicial map from K to Δ^n defined on the vertices by $v \mapsto e_{\lambda(v)}$. Consider the induced map λ_* on the $\mathbb{Z}_2$-simplicial chain complex level

$$\lambda_* : C_*(K; \mathbb{Z}_2) \to C_*(\Delta^n; \mathbb{Z}_2).$$

Let o denote the element of $C_n(K; \mathbb{Z}_2)$ given by the sum of all n-simplices of K. Clearly, the Sperner lemma holds if $\lambda_n(o) = \sigma^n$, the generator (and only nontrivial element) of $C_n(\Delta^n; \mathbb{Z}_2)$. Now consider the following diagram, which is commutative by the fact that λ_* is a chain map:

$$
\begin{array}{ccc}
C_n(K; \mathbb{Z}_2) & \xrightarrow{\lambda_n} & C_n(\Delta^n; \mathbb{Z}_2) \\
{\scriptstyle \partial_n}\downarrow & & \downarrow{\scriptstyle \partial_n} \\
C_{n-1}(K; \mathbb{Z}_2) & \xrightarrow{\lambda_{n-1}} & C_{n-1}(\Delta^n; \mathbb{Z}_2)
\end{array}
$$

Hence, it suffices to show that $\lambda_{n-1}\partial_n(o) \neq 0$. In order to compute $\lambda_{n-1}\partial_n(o)$, let $\tau_1, \ldots, \tau_{n+1}$ denote the $(n-1)$-dimensional faces of Δ^n. Define $c_i \in C_{n-1}(K; \mathbb{Z}_2)$ to be the sum of all $(n-1)$-dimensional faces of K that lie in τ_i. Then $\partial_n(o) = \sum_{i=1}^{n+1} c_i$, and by the induction hypothesis, $\lambda_{n-1}(c_i) = \tau_i$, and hence $\lambda_{n-1}\partial_n(o) = \sum_{i=1}^{n+1} \tau_i \neq 0$. □

Brouwer's Theorem via Sperner's Lemma

Finally, we can give an elementary proof of Brouwer's fixed-point theorem relying on Sperner's lemma.

Proof (of Brouwer's fixed-point theorem). Since $\mathbb{B}^n$ and the standard n-simplex are homeomorphic, we may consider a continuous map $f : |\Delta^n| \to |\Delta^n|$, where Δ^n is the (geometric) simplicial complex given by the standard n-simplex and all its faces. Consider the kth barycentric subdivisions $\text{sd}^k \Delta^n$, $k \geq 1$. If, for some k, one of the vertices of $\text{sd}^k \Delta^n$ happens to be a fixed point, we are done. Otherwise, we construct a sequence $(\sigma_k)_{k \geq 1}$ of simplices of decreasing size such that any *accumulation point* of this sequence will be a fixed point of f. By an accumulation point we mean a point $x \in |\Delta^n|$ such that each ε-ball about x contains infinitely many of the σ_k, $k \geq 1$. In order to do this, we endow the kth barycentric subdivision $\text{sd}^k \Delta^n$ with a Sperner labeling as follows. For $v \in \text{vert}(\text{sd}^k \Delta^n)$ let $\lambda(v)$ be the smallest i such that the ith coordinate of $f(v) - v$ is negative, i.e.,

$$\lambda(v) = \min\{i : f(v)_i - v_i < 0\}.$$

Such an i exists, since the sum over all coordinates of $f(v) - v$ is zero and v is not a fixed point. This labeling is indeed a Sperner labeling, since for $v_i = 0$, we certainly have $f(v)_i - v_i \geq 0$. Hence, by Sperner's lemma, there exists a fully labeled simplex σ_k.

Now let x be an accumulation point of the sequence (σ_k) of simplices. For the existence of such an x we refer to Corollary B.48 and Exercise 16 on page 195. Hence, for each i and any $\varepsilon > 0$, there exist a $k \geq 1$ and a vertex $v \in \text{vert}(\sigma_k)$ such that $|x - v| < \varepsilon$ and $f(v)_i - v_i < 0$. By continuity, we obtain the inequality $f(x)_i - x_i \leq 0$ for all i. But since the sum $\sum_{i=1}^{n+1}(f(x)_i - x_i)$ is zero, this is possible only if $f(x) = x$. □

Brouwer's Theorem via Homology Theory

As previously announced, we provide a proof of Brouwer's theorem using only the basics of homology theory typically taught in a first course on algebraic topology. More details on the necessary background can be found in Appendix B.

Proof (using homology theory). Assume that $f : \mathbb{B}^n \to \mathbb{B}^n$ is a continuous map without a fixed point. For each x, consider the ray from $f(x)$ in the direction of x. This ray hits the boundary sphere $\mathbb{S}^{n-1} = \{x \in \mathbb{R}^n : \|x\| = 1\}$ of $\mathbb{B}^n$ in a point that we call $r(x)$; see Fig. 1.6.

Then $r : \mathbb{B}^n \to \mathbb{S}^{n-1}$ is a continuous map that when restricted to the sphere, is the identity map, i.e., $r \circ i = \text{id}_{\mathbb{S}^{n-1}}$, where $i : \mathbb{S}^{n-1} \hookrightarrow \mathbb{B}^n$ is the inclusion map. Such a map is called a retraction map. We obtain the following induced maps in homology:

$$\tilde{H}^{n-1}(\mathbb{S}^{n-1}; \mathbb{Z}) \xrightarrow{i_*} \tilde{H}^{n-1}(\mathbb{B}^n; \mathbb{Z}) \xrightarrow{r_*} \tilde{H}^{n-1}(\mathbb{S}^{n-1}; \mathbb{Z}).$$

Now, $r_* \circ i_* = (r \circ i)_* = (\text{id}_{\mathbb{S}^{n-1}})_* = \text{id}_{\tilde{H}^{n-1}(\mathbb{S}^{n-1}; \mathbb{Z})}$ is the identity map of $\tilde{H}^{n-1}(\mathbb{S}^{n-1}; \mathbb{Z})$, which is isomorphic to the integers $\mathbb{Z}$. But since $\tilde{H}^{n-1}(\mathbb{B}^n; \mathbb{Z})$ is trivial, we arrive at a contradiction. □

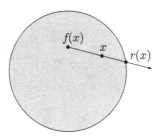

Fig. 1.6 The retraction map r

Sperner's Lemma Derived from Brouwer's Theorem

As remarked in the introduction to this chapter, Sperner's lemma may be considered a combinatorial analogue of Brouwer's theorem. This is due to the fact that there is also a way to deduce Sperner's lemma from Brouwer's fixed-point theorem. We end this section by proving this.

Proof (of Sperner's lemma with Brouwer's fixed-point theorem). Let $\lambda : \mathrm{vert}(K) \to [n+1]$ be a Sperner labeling of a triangulation K of Δ^n. We construct a continuous map $f : |\Delta^n| \to |\Delta^n|$ as an affine linear extension of the simplicial map from K to Δ^n defined on the vertices of K by $v \mapsto e_{\lambda(v)+1(\bmod n+1)}$.

Observe that there exists a fully labeled simplex if and only if f is surjective. To prove this, it suffices to show that some point in the interior of $|\Delta^n|$ is in the image of f. It is a good exercise to show that the map f is fixed-point-free on the boundary of $|\Delta^n|$. The existence of a fixed point yields the desired conclusion. □

1.2 Envy-Free Fair Division

Sometimes it is hard to divide a piece of cake among several people, especially if the cake contains tasty ingredients such as nuts, raisins, and chocolate chips that may be distributed unevenly and if we take into account that preferences among several people are often quite different. This calls for a procedure to find a solution that is satisfying for everyone.

In order to do this, we first have to specify more precisely what we mean by "satisfying for everyone." Ludwig Erhard, German chancellor in the 1960s, once said, "Compromise is the art of dividing a cake such that everyone is of the opinion he has received the largest piece." This seemingly paradoxical statement is pretty much what our definition of envy-free fair division is going to be!

Fig. 1.7 A cake with different tasty ingredients

A Fair-Division Model

Let's say we have n people among whom the cake is to be divided. Each person might have his or her own idea about which content of the cake is valuable: for one it's the nuts, for another it's the chocolate, and so on. Figure 1.7 shows a cake with colored chocolate beans indicating the different preferences. We model the piece of cake with an interval $I = [0, 1]$ (which might be thought of as a projection of the cake), and the predilections of the people by *continuous probability measures* $\mu_1, \ldots, \mu_n$. Continuous means that the functions $t \mapsto \mu_i\big([0, t)\big)$ are continuous in t. The continuity condition implies that the measures evaluate to zero on single points, i.e., $\mu_i(\{t\}) = 0$ for all $t \in I$.

Let's assume that the cake is divided into n pieces (each measurable with respect to all μ_i), i.e., $I = A_1 \cup \cdots \cup A_n$, where $A_i \cap A_j$ is a finite set of points for each $i \neq j$, and person i is to receive the piece $A_{\pi(i)}$ for some permutation π.

Definition 1.4. The division $(A_1, \ldots, A_n; \pi)$ of the cake is called *fair* if $\mu_i(A_{\pi(i)}) \geq \frac{1}{n}$ for all i. It is called *envy-free* if $\mu_i(A_{\pi(i)}) \geq \mu_i(A_{\pi(j)})$ for all i, j.

The last condition says that each person receives a piece that is (according to its measure) at least as large as all the other pieces.

The permutation π might seem unnecessary at this point, but for the purpose of upcoming proofs we need to be able to assign the n pieces of a fixed division to the n people.

Practical Cake-Cutting

There are several interesting approaches to obtaining solutions to *fair cake-cutting*. The simplest is the following. Let t be the smallest value such that there exists an i with $\mu_i([0, t]) = \frac{1}{n}$. This means that for all other j, the rest of the cake, i.e., $[t, 1]$, has size at least $\frac{n-1}{n}$. Therefore, person i is to receive the piece $[0, t]$ and the others proceed by induction on the rescaled piece. Note that this procedure produces $n - 1$ cuts of the cake, and hence n intervals, i.e., the fewest number possible.

This procedure is often referred to as the *moving-knife algorithm,* and it works as follows. Some person slowly moves a knife along the cake. If the portion of the cake that has been covered by the moving knife has reached size $\frac{1}{n}$ for some person, then this person yells "stop!" The cake is cut right there, and the person who yelled receives the piece. If more than one person yelled, the piece is given to one of them. From a practical viewpoint, this has the advantage that every person feels treated fairly. But of course, in general, the divisions obtained in this way are not envy-free. There are several algorithms one can use to obtain an envy-free division of the cake. But these often require many cuts of the cake; cf. [RW98].

The Simplex as Solution Space

Here we want to concentrate on the existence and approximation of an envy-free solution that can be obtained by $n - 1$ cuts.

A division of the unit interval into n successive intervals is determined by the vector $(t_1, \ldots, t_n)$ of their lengths, and all possible such division vectors constitute the standard $(n - 1)$-simplex if we allow intervals of length zero.

The first proof of the existence of an envy-free fair-division solution with $n - 1$ cuts by Woodall [Woo80] is a construction whose topological engine is Brouwer's fixed-point theorem. There is an easier construction that—not surprisingly—relies on Sperner's lemma. Moreover, this construction yields a method to find approximate solutions described by Su [Su99], which has a nice implementation called the "Fair division calculator" and is available on the Internet.

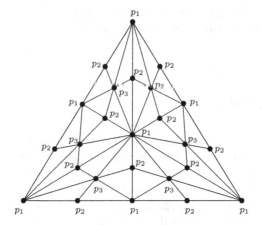

Fig. 1.8 First labeling of vert(sd^k Δ^{n-1})

A Sperner Labeling Approach

In order to find an approximate solution, consider a barycentric subdivision sd^k Δ^{n-1} for some k. Our construction will consist of two consecutive labelings of the vertices, the second of which is going to be a Sperner labeling. The first labeling is rather simple. Label the vertices of sd^k Δ^{n-1} with labels $p_1, \ldots, p_n$ in such a manner that the vertices of each $(n-1)$-simplex obtain all labels $p_1, \ldots, p_n$, as demonstrated in Fig. 1.8. Such a labeling is easy to derive and is the content of Exercise 4.

To define the second labeling, consider a vertex $v = (t_1, \ldots, t_n)$ of sd^k Δ^{n-1} with, say, label p_{i_0}. It defines the division $I = [0, t_1] \cup [t_1, t_1 + t_2] \cup \cdots \cup [t_1 + \cdots + t_{n-1}, 1]$. Denote the kth interval by I_k and let $\mu(v) = \max\{\mu_{i_0}(I_1), \ldots, \mu_{i_0}(I_n)\}$ be the size of the largest piece according to person i_0. Define the labeling by

$$\lambda(v) = \min\{j : \mu(I_j) = \mu(v)\}.$$

In other words, $\lambda(v)$ describes the number of a piece that is largest for person i_0. Certainly

$$\lambda(v) \in \{j : t_j \neq 0, j \in [n]\},$$

since the largest piece will not be an interval of length 0, and hence λ is a bona fide Sperner labeling. For an example see Fig. 1.9.

By Sperner's lemma, we obtain a fully labeled $(n-1)$-simplex σ_k, which means that the n different people associated with the n vertices of σ_k all choose a different interval. More precisely, σ_k defines a permutation $\pi_k : [n] \rightarrow [n]$, where $\pi_k(i) = j$ if $\lambda(v) = j$ for the vertex v of σ_k labeled with p_i. Now let x_k be the barycenter

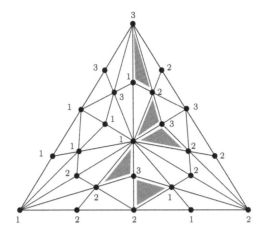

Fig. 1.9 An example of the labeling λ

of the simplex σ_k and consider the sequence (x_k). By compactness, there exists a convergent subsequence (x_{i_k}). Since there is only a finite number of permutations of $[n]$, we may even choose this subsequence with the additional property that the sequence (π_{i_k}) of associated permutations is constant. Let $x = (t_1, \ldots, t_n)$ be the limit of this subsequence and π the constant permutation. The associated division $(A_1, \ldots, A_n; \pi)$ is the desired envy-free solution, as is easy to prove and is the content of an exercise. Thus we obtain the following result.

Theorem 1.5. *Let $\mu_1, \ldots, \mu_n$ be n continuous probability measures on the unit interval. Then there exists an envy-free division $(A_1, \ldots, A_n; \pi)$ such that all of the A_i are intervals.* $\qquad\Box$

Note that, moreover, an approximate solution can be found in a finite number of steps: in fact, for any given $\varepsilon > 0$, there exist a $k \geq 0$, a simplex $\sigma_k \in \operatorname{sd}^k \Delta^{n-1}$, and a permutation π with the property that the division associated with the barycenter of σ_k together with π is envy-free up to an error of ε.

1.3 The Borsuk–Ulam Theorem and Tucker's Lemma

The Borsuk–Ulam theorem is a classical theorem in algebraic topology, and next to Brouwer's theorem, is one of the main results typically proven in an algebraic topology course to show the power of homological methods. For some historical background on Stan Ulam and the history of the theorem, I recommend Gian-Carlo Rota's wonderful article [Rot87]. The Borsuk–Ulam theorem is often illustrated by the claim that at any moment in time, there is a pair of antipodal points on the surface of the earth with the same air pressure and temperature. We will present

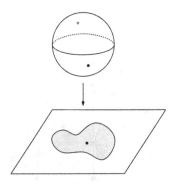

Fig. 1.10 A sphere made flat

four versions of the theorem, most of which will play some role in the sequel. An illustration of the third version is given in Fig. 1.10.

Theorem 1.6 (Borsuk–Ulam). *The following statements hold.*

1. *If $f : \mathbb{S}^n \to \mathbb{S}^m$ is a continuous antipodal map, i.e., $f(-x) = -f(x)$ for all $x \in \mathbb{S}^n$, then $n \leq m$.*
2. *If $f : \mathbb{S}^n \to \mathbb{R}^n$ is a continuous antipodal map, then there exists an $x \in \mathbb{S}^n$ such that $f(x) = 0$.*
3. *If $f : \mathbb{S}^n \to \mathbb{R}^n$ is a continuous map, then there exists $x \in \mathbb{S}^n$ such that $f(x) = f(-x)$.*
4. *If $\mathbb{S}^n$ is covered by $n + 1$ subsets $S_1, \ldots, S_{n+1}$ such that each of $S_1, \ldots, S_n$ is open or closed, then one of the sets contains an antipodal pair of points, i.e., there exist an $i \in [n + 1]$ and $x \in \mathbb{S}^n$ such that $x, -x \in S_i$.*

Since Brouwer's fixed-point theorem is intimately related to Sperner's lemma, the same is true for the Borsuk–Ulam theorem and a lemma by Tucker. In the sequel, we will present a proof of the Borsuk–Ulam theorem by means of Tucker's lemma. For a proof using standard methods from algebraic topology, I recommend Bredon [Bre93]. But in order to get used to the different ways the Borsuk–Ulam theorem was stated, we will show that each of the four versions easily implies the others.

Proof (of the equivalences). $(1 \Rightarrow 2)$ Assume $f : \mathbb{S}^n \to \mathbb{R}^n$ is an antipodal map without zero. Then the map

$$x \longmapsto \frac{f(x)}{\|f(x)\|}$$

is (by compactness of $\mathbb{S}^n$) a continuous antipodal map from $\mathbb{S}^n$ to $\mathbb{S}^{n-1}$, contradicting 1.

$(2 \Rightarrow 3)$ Let $f : \mathbb{S}^n \to \mathbb{R}^n$ be a continuous map. Consider the continuous *and antipodal* map $g : \mathbb{S}^n \to \mathbb{R}^n$ defined by $g(x) = f(x) - f(-x)$. By statement 2., g has a zero, x, which yields the desired property for f.

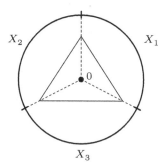

Fig. 1.11 A cover of the $(n-1)$-sphere with $n+1$ closed sets

$(3 \Rightarrow 4)$ Let $\mathbb{S}^n$ be covered by $n+1$ subsets $S_1, \ldots, S_{n+1}$, such that each of $S_1, \ldots, S_n$ is open or closed. Assume that none of $S_1, \ldots, S_n$ contains an antipodal pair of points. Then $x \in S_i$ implies $\mathrm{dist}(-x, S_i) > 0$ for each $i \in [n]$ and $x \in \mathbb{S}^n$. We show this by considering separately the cases in which S_i is closed or open. If $A = S_i$ is closed and $x \in A$, then $-x \notin A$, and therefore $\mathrm{dist}(-x, A) > 0$. If $U = S_i$ is open and $x \in U$, then $\mathrm{dist}(x, \mathbb{S}^n \setminus U) > 0$, and since $-U \subseteq \mathbb{S}^n \setminus U$, we derive $\mathrm{dist}(-x, U) = \mathrm{dist}(x, -U) \geq \mathrm{dist}(x, \mathbb{S}^n \setminus U) > 0$.

We will now find an antipodal pair of points in S_{n+1} as follows. Consider the continuous map

$$f : \mathbb{S}^n \longrightarrow \mathbb{R}^n,$$

$$x \longmapsto \begin{pmatrix} \mathrm{dist}(x, S_1) \\ \vdots \\ \mathrm{dist}(x, S_n) \end{pmatrix}.$$

By assumption there exists an $x \in \mathbb{S}^n$ with $f(x) = f(-x)$. We are done if we can show that $x, -x \notin S_1 \cup \cdots \cup S_n$. We check this by showing $x, -x \notin S_i$ for each $i \in [n]$. If $\mathrm{dist}(x, S_i) = \mathrm{dist}(-x, S_i) > 0$, then clearly $x, -x \notin S_i$. If $\mathrm{dist}(x, S_i) = \mathrm{dist}(-x, S_i) = 0$, then $x, -x \notin S_i$ by the discussion above.

$(4 \Rightarrow 1)$ Assume that there is an antipodal map $f : \mathbb{S}^n \to \mathbb{S}^{n-1}$. Now the important observation is that the $(n-1)$-dimensional sphere can be covered with $n+1$ closed sets, none of which contains an antipodal pair. To see this, consider an n-simplex in $\mathbb{R}^n$ with 0 in the interior. The radial projections $X_1, \ldots, X_{n+1}$ of the $n+1$ facets of dimension $n-1$ to the sphere yield the desired cover, as demonstrated in Fig. 1.11. Now let $S_i = f^{-1}(X_i)$. By continuity of f, the S_i are closed, and by the antipodality of f, they do not contain any antipodal pair of points. $\square$

For the last implication, note that the family, $X_1, \ldots, X_{n+1}$, of open sets $X_i = \{x \in \mathbb{S}^{n-1} \subseteq \mathbb{R}^n : x_i > 0\}$ for $i \in [n]$ and the closed set $X_{n+1} = \{x \in \mathbb{S}^{n-1} \subseteq \mathbb{R}^n : \forall i \in [n] : x_i \leq 0\}$ would have worked as well.

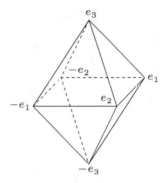

Fig. 1.12 The cross polytope Q^3

Tucker's Lemma

To formulate Tucker's lemma we will be interested in subdivisions of the n-dimensional sphere $\mathbb{S}^n$ that refine the "triangulation" of $\mathbb{S}^n$ by the coordinate hyperplanes. More precisely, we will consider subdivisions K of the boundary complex Γ^n of the cross polytope, which we will introduce now.

The $(n+1)$-dimensional *cross polytope* is defined to be the convex hull $Q^{n+1} = \text{conv}(\{\pm e_1, \ldots, \pm e_{n+1}\})$. Its boundary is the polyhedron of a geometric simplicial complex, whose simplices are given by the convex hulls of sets $\{\varepsilon_{i_1} e_{i_1}, \ldots, \varepsilon_{i_k} e_{i_k}\}$, not containing an antipodal pair $\pm e_j$. We denote the (geometric) simplicial complex given by this collection of geometric simplices by

$$\Gamma^n = \big\{ \text{conv}(\{\varepsilon_{i_1} e_{i_1}, \ldots, \varepsilon_{i_k} e_{i_k}\}) : 0 \le k \le n+1,$$
$$1 \le i_1 < \cdots < i_k \le n+1, \varepsilon_{i_j} \in \{\pm 1\}\big\}.$$

An alternative way to construct Γ^n is to take the $(n+1)$-fold join of two-point sets, i.e., 0-spheres

$$\Gamma^n = \{\pm e_1\} * \{\pm e_2\} * \cdots * \{\pm e_{n+1}\},$$

where $\{\pm e_i\}$ serves as an abbreviation of the geometric complex $\{\emptyset, +e_1, -e_1\}$. An illustration is given in Fig. 1.12.

Tucker's lemma is concerned with *antipodally symmetric triangulations* K, i.e., triangulations with the property that $\sigma \in K$ if and only if $-\sigma \in K$.

Lemma 1.7 (Tucker). *Let K be an antipodally symmetric subdivision of Γ^n, and $\lambda : \text{vert}(K) \to \{\pm 1, \ldots, \pm n\}$ an antipodally symmetric labeling of K, i.e., $\lambda(-v) = -\lambda(v)$ for all v. Then there exists a complementary edge, i.e., an edge $uv \in K$ with $\lambda(u) + \lambda(v) = 0$.*

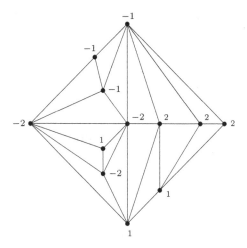

Fig. 1.13 A front side of a labeled subdivision of Γ^2

Since the triangulation K and the labeling are antipodally symmetric, we can sketch an example by just showing one side of the boundary of the cross polytope; cf. Fig. 1.13.

Tucker's lemma is an immediate corollary of the Borsuk–Ulam theorem. The proof of this is the content of Exercise 8. Conversely, the Borsuk–Ulam theorem may be derived from Tucker's lemma, as we show next. Afterwards we will be concerned with a combinatorial proof of Tucker's lemma that also finds a complementary edge, in a manner analogous to the combinatorial proof of Sperner's lemma. The first such proof, given by Freund and Todd [FT81], even proves that there is always an odd number of complementary edges. We will present a more recent proof by Prescott and Su, and moreover give an elegant direct proof for the existence of a complementary edge.

Proof (of the Borsuk–Ulam theorem with the Tucker lemma). Since $|\Gamma^n|$ and $\mathbb{S}^n$ are homeomorphic, we may assume for contradiction that there exists an antipodal map $f : |\Gamma^n| \rightarrow \mathbb{R}^n$ without a zero. Hence, there exists an $\varepsilon > 0$ such that $\|f\|_\infty \geq \varepsilon$, i.e., for each $x \in \mathbb{S}^n$ there exists a coordinate i with $|f_i(x)| \geq \varepsilon$. By continuity of f, there exists a k such that for all edges uv of $K = \mathrm{sd}^k \Gamma^n$, we have $\|f(u) - f(v)\|_\infty < \varepsilon$. We construct an antipodally symmetric labeling $\lambda : \mathrm{vert}(K) \rightarrow \{\pm 1, \ldots, \pm n\}$ as follows. Let

$$i(v) = \min\{i : |f_i(v)| \geq \varepsilon\},$$

and define

$$\lambda(v) = \begin{cases} +i(v) & \text{if } f_{i(v)}(v) \geq \varepsilon, \\ -i(v) & \text{if } f_{i(v)}(v) \leq -\varepsilon. \end{cases}$$

Now $|f_i(-v)| = |-f_i(v)| = |f_i(v)|$ implies $i(-v) = i(v)$, and hence $\lambda(-v) = -\lambda(v)$. By Tucker's lemma, there exists an edge uv in K such that for some $i \in [n]$, we have $\lambda(u) = +i$ and $\lambda(v) = -i$ (after maybe switching u and v). Hence, by definition of the labeling, $f_i(u) \geq \varepsilon$ and $f_i(v) \leq -\varepsilon$, contradicting $\|f(u) - f(v)\|_\infty < \varepsilon$.   □

1.4 A Generalization of Tucker's Lemma

As announced, we now turn our attention to combinatorics and consider a generalization of Tucker's lemma by Ky Fan [Fan52]. The idea is to decouple the size of the label set from the dimension n. We will be interested in certain *alternatingly* labeled simplices. Let K be an antipodally symmetric subdivision of Γ^n and $\lambda : \mathrm{vert}(K) \to \{\pm 1, \ldots, \pm m\}$ an antipodally symmetric labeling of K. A d-simplex σ is called $+$-*alternating*, resp. $-$-*alternating*, if it has labels $\{+j_0, -j_1, +j_2, \ldots, (-1)^d j_d\}$, resp. $\{-j_0, +j_1, -j_2, \ldots, (-1)^{d+1} j_d\}$, where $1 \leq j_0 < j_1 < \cdots < j_d \leq m$. For an illustration we refer to Fig. 1.14.

Theorem 1.8 (Ky Fan, weak version). *Let K be an antipodally symmetric subdivision of Γ^n and $\lambda : \mathrm{vert}(K) \to \{\pm 1, \ldots, \pm m\}$ an antipodally symmetric labeling of K without complementary edges. Then there exists an odd number of $+$-alternating n-simplices.*

Corollary 1.9 (Tucker's lemma). *The existence of an alternatingly labeled simplex of dimension n implies $m \geq n + 1$, and hence any antipodally symmetric labeling $\mathrm{vert}(K) \to \{\pm 1, \ldots, \pm n\}$ must contain a complementary edge.*   □

The first proof of Ky Fan's theorem that we present is due to Prescott and Su [PS05] and requires a generalization of the concept of an alternating simplex.

A d-simplex σ is called ε-*almost-alternating* if it is not alternating but has a $(d-1)$-face, i.e., a facet, that is ε-alternating for some $\varepsilon \in \{\pm\}$. It is easy to see that there is indeed no ambiguity in ε, in other words, every ε-almost-alternating simplex without a complementary edge has exactly two facets that are ε-alternating. Figure 1.15 shows the different possibilities.

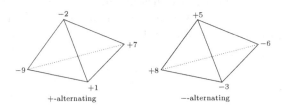

Fig. 1.14 A $+$- and a $-$-alternating 3-dimensional simplex

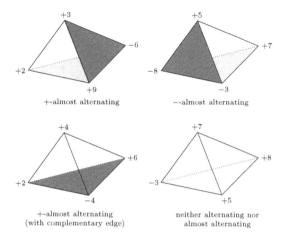

+-almost alternating   --almost alternating

+-almost alternating   neither alternating nor
(with complementary edge)   almost alternating

Fig. 1.15 A 3-dimensional simplex with various labelings

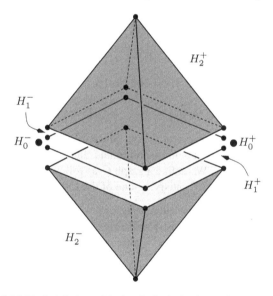

Fig. 1.16 Exploded view of the hemispherical subcomplexes for $n = 2$

Now consider the following family of hemispherical subcomplexes of Γ^n, illustrated in Fig. 1.16. The respective north and south hemispheres will obtain a sign $\varepsilon \in \{\pm\}$:

$$H_0^\varepsilon = \{\varepsilon e_1\},$$
$$H_1^\varepsilon = \{\pm e_1\} * \{\varepsilon e_2\},$$

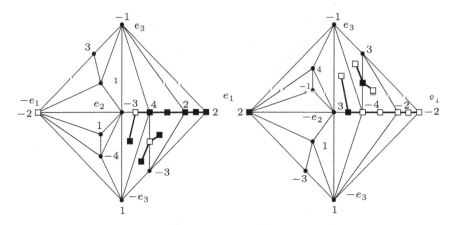

Fig. 1.17 Front and back sides of a labeling and the graph G for $n = 2$ and $m = 4$

$$H_2^\varepsilon = \{\pm e_1\} * \{\pm e_2\} * \{\varepsilon e_3\},$$

$$\vdots$$

$$H_n^\varepsilon = \{\pm e_1\} * \cdots * \{\pm e_n\} * \{\varepsilon e_{n+1}\}.$$

Proof (of Ky Fan's theorem constructively). Since K is a subdivision of Γ^n, for each simplex σ of K there exist a unique minimal i and $\varepsilon \in \{\pm\}$ such that $\sigma \in H_i^\varepsilon$, the *carrier hemisphere of* σ. We define a graph G as follows. The vertices of G are those simplices σ carried by H_d^ε satisfying one of the following:

* σ has dimension $d - 1$ and is ε-alternating,
* σ has dimension d and is ε-almost-alternating,
* σ is d-dimensional and either $+$- or $-$-alternating.

Two vertices, σ and τ, constitute an edge if (up to switching roles of σ and τ):

* σ is a facet of τ, and
* If τ is carried by H_k^ε, then σ is ε-alternating.

An example of such a graph G is shown in Fig. 1.17. The vertices of G corresponding to $+$-(almost-) alternating simplices are indicated by a black box, while the vertices corresponding to $-$-(almost-) alternating simplices are indicated by a white box.

The graph has vertex degrees 1 and 2 only, as we will see now. An ε-alternating $(d - 1)$-simplex σ carried by H_d^ε is a facet of exactly two d-simplices carried by the same hemisphere. Each of these simplices is either ε-alternating or ε-almost-alternating. And hence σ has degree 2.

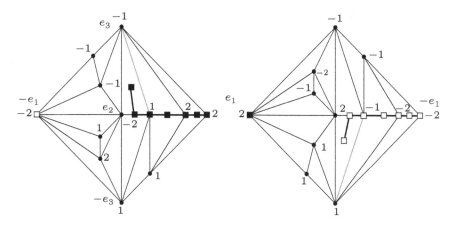

Fig. 1.18 How to find a complementary edge

An ε-almost-alternating d-simplex τ carried by H_d^ε has exactly two ε-alternating $(d-1)$-simplices in the boundary, and hence τ has degree 2.

Finally, let's consider a d-dimensional ε'-alternating simplex σ carried by H_d^ε. Unless $d = 0$, σ has exactly one facet that is ε-alternating obtained by deleting the vertex with the label of smallest or largest absolute value. And unless $d = n$, exactly one of the $(d+1)$-simplices containing σ is contained in H_{d+1}^ε.

Note that the graph G is invariant under the antipodal action on K and hence consists of antipodal pairs of paths and cycles. The endpoints of the paths, i.e., the only vertices of degree 1, are $\{\pm e_1\} = H_0^\pm$ and the alternating simplices of dimension n. It is an easy exercise to see that the two vertices $\{\pm e_1\}$ will not be connected by a path. Hence the total number of alternating n-simplices will be twice an odd number, half of which are $+$-alternating. □

Observe that the construction of the graph G also works in the presence of complementary labeled edges. In this case, some of the ε-almost-alternating simplices will contain such edges. These almost-alternating simplices will have only one ε-alternating facet. And since there are insufficient labels to obtain an alternating n-simplex, the constructed path from $H_0^\pm$ will reach an almost-alternating simplex containing a complementary edge. Hence we have found a method to easily detect a complementary edge in the situation of Tucker's lemma in which we have a labeling $\mathrm{vert}(K) \to \{\pm 1, \ldots, \pm n\}$. For an illustration see Fig. 1.18.

Note that this procedure does not find, nor match up, all simplices containing a complementary edge. But in the solution of the consensus $\frac{1}{2}$-division problem covered in the next section, it suffices to find one complementary edge.

We will now turn our attention to a stronger form of Ky Fan's theorem that resembles a Stokes formula in a combinatorial setting. Note that there are no symmetry conditions on the subdivision or on the labeling.

Theorem 1.10 (Ky Fan, strong version).   *Let K be a subdivision of the upper hemisphere H_n^+ of Γ^n and λ : vert(K) $\to$ $\{\pm 1, \dots, \pm m\}$ a labeling of K without complementary edges. Then the number of $+$-alternating n-simplices plus the number of $-$-alternating n-simplices is congruent modulo 2 to the number of $+$-alternating $(n - 1)$-simplices in the boundary $\partial H_n^+ = H_n^+ \cap H_n^-$.*

The idea behind the following proof is due to Frédéric Meunier [Meu05].

Proof. Consider the chain complex $(C_*(K; \mathbb{Z}_2), \partial)$ of K with $\mathbb{Z}_2$-coefficients. We define the following (mod 2)-counting function:

$$\beta : C_{n-1}(K; \mathbb{Z}_2) \to \mathbb{Z}_2,$$

$$\sum_{\tau \in T} \tau \mapsto |\{\tau : \tau \in T, \tau \text{ is } +\text{-alternating}\}| \quad \text{mod } 2.$$

Now let $\alpha : C_n(K; \mathbb{Z}_2) \to \mathbb{Z}_2$ be defined by $\alpha = \beta \circ \partial$. Miraculously, α turns out to be the right counting function in dimension n, i.e.,

$$\alpha \left(\sum_{\sigma \in S} \sigma \right) = \beta \left(\partial \sum_{\sigma \in S} \sigma \right) g = |\{\sigma : \sigma \in S, \sigma \text{ is } +\text{- or } -\text{-alternating}\}| \quad \text{mod } 2.$$

In order to check this, note that for an n-dimensional simplex σ, $\alpha(\sigma)$ may be nonzero only if σ is an alternating or almost-alternating simplex. The verification that $\alpha(\sigma)$ is zero modulo 2 in the case that σ is almost-alternating requires the nonexistence of complementary edges.

Now let $c = \sum_{\sigma \in K, \dim(\sigma)=n} \sigma$ be the sum of all n-simplices in K. Then $\partial C = \sum_{\tau \in K, \dim(\tau)=n-1, \tau \subset \partial H_n^+} \tau$. Now the equality $\alpha(c) = \beta(\partial c)$ yields the result.   □

The strong Ky Fan theorem implies the weak version as follows.

Proof (Strong implies weak Ky Fan). We proceed by induction on the dimension. For $n = 0$, the statement is obvious. Now assume $n > 0$. Let $\alpha_n^{\pm}$ be the number of $\pm$-alternating n-simplices of K lying in H_n^+, and let β_{n-1} be the number of $+$-alternating $(n - 1)$-simplices in K lying in $\partial H_n^+ = H_n^+ \cap H_n^-$ for $i = 1, \dots, n - 1$. By the induction hypothesis, $\beta_{n-1} \equiv 1 \mod 2$. Note that by the antipodal symmetry of the labeling, the number of $+$-alternating n-simplices in K is equal to $\alpha_n^+ + \alpha_n^-$. The strong Ky Fan theorem says that $\alpha_n^+ + \alpha_n^- \equiv \beta_{n-1} \mod 2$, and hence we are done.   □

Here we witness an amazing phenomenon that is surprisingly common in mathematics: we have a much simpler proof for an even stronger result. In particular, this frequently happens with statements that admit an inductive proof. In these cases, it is sometimes important to have a statement strong enough to get the induction to work. But note that our last proof no longer yields a procedure to find a complementary edge.

The strong Ky Fan theorem can easily be generalized to a wider class of spaces, i.e., pseudomanifolds with boundary (see Exercise 13 on page 33). Moreover, the proof of the strong Ky Fan theorem that we presented can be interpreted in a more general framework that sheds light on a method for obtaining such formulas in general.

1.5 Consensus $\frac{1}{2}$-Division

Assume that two daughters have inherited a piece of land and they are to divide the land between them. In order to keep the family peace, it is necessary that all family members have the impression that the share of the land each daughter obtains is worth exactly half the total value. One of many similar real-world applications mentioned in the literature is the Law of the Sea Treaty [SS03]. We will now model the situation by assuming that the piece of land is an interval endowed with probability measures for the individual preferences of the family members.

Let $\mu_1, \ldots, \mu_n$ be n continuous probability measures on the unit interval $I = [0, 1]$. We are searching for a subdivision $I = A_{+1} \cup A_{-1}$ such that $\mu_i(A_{+1}) = \mu_i(A_{-1}) = \frac{1}{2}$ for all i. Note that this implies $\mu_i(A_{+1} \cap A_{-1}) = 0$ for all i. Of course $A_{\pm 1}$ need only be two measurable sets, but we are interested in particularly nice solutions, i.e., we want $A_{\pm 1}$ to be a union of finitely many intervals each. Observe that since the n measures might have disjoint support, in general we will need to make at least n cuts in order to divide each measure in half, and hence divide I into $n + 1$ intervals that are to be distributed into the two sets $A_{\pm 1}$.

The Cross Polytope as Solution Space

Let's consider the space X_n of all such divisions $I = A_{+1} \cup A_{-1}$ that arise by cutting the interval n times and assigning each interval to exactly one of A_{+1} and A_{-1}:

$$X_n = \left\{ ((\varepsilon_0, t_0), \ldots, (\varepsilon_n, t_n)) : \varepsilon_i \in \{\pm 1\}, t_i \geq 0, \sum_{i=0}^{n} t_i = 1 \right\}.$$

Here t_i yields the length of the $(i+1)$st interval and ε_i indicates whether the interval is assigned to A_{+1} or A_{-1}. For example, the vector $x = ((-1, \frac{1}{3}), (+1, \frac{1}{2}), (-1, \frac{1}{6}))$ corresponds to the division of I into $A_{-1}(x) = [0, \frac{1}{3}] \cup [\frac{5}{6}, 1]$ and $A_{+1}(x) = [\frac{1}{3}, \frac{5}{6}]$.

Let's take a closer look at X_n. If $((\varepsilon_0, t_0), \ldots, (\varepsilon_n, t_n)) \in X_n$ has the property that $t_j = 0$ for some fixed j, then the sign ε_j is irrelevant, since the interval of length $t_j = 0$ has measure zero with respect to any of the μ_i. We hence introduce the equivalence relation $\sim$ on X_n given by

$$\big((\varepsilon_0, t_0), \ldots, (+1, t_j), \ldots, (\varepsilon_n, t_n)\big) \sim \big((\varepsilon_0, t_0), \ldots, (-1, t_j), \ldots, (\varepsilon_n, t_n)\big)$$

if and only if $t_j = 0$. The quotient space $X_n/\!\!\sim$ of all possible solutions turns out to be an object we know quite well: it is the (polyhedron of the) $(n+1)$-fold join of a two-point space with elements $\{\pm 1\}$, which we identify with the multiplicative two-element group $\mathbb{Z}_2$. This space, $\mathbb{Z}_2^{*(n+1)}$, is nothing but the boundary complex of the $(n+1)$-dimensional cross polytope. In particular, it is a bona fide n-sphere. The group $\mathbb{Z}_2$ acts on $X_n/\!\!\sim$ via multiplication in all factors of the join, i.e.,

$$\varepsilon \cdot \big((\varepsilon_0, t_0), \ldots, (\varepsilon_n, t_n)\big) = \big((\varepsilon \cdot \varepsilon_0, t_0), \ldots, (\varepsilon \cdot \varepsilon_n, t_n)\big),$$

which corresponds to the antipodal action on Γ^n. Note that the subdivision corresponding to the antipode $-x$ of x arises from the subdivision corresponding to x by interchanging A_{+1} with A_{-1}, i.e., $A_{\pm 1}(-x) = A_{\mp 1}(x)$.

The existence of a solution to the consensus $\frac{1}{2}$-division problem now immediately follows from the Borsuk–Ulam theorem. The reader might want to think about it before we present the solution.

Define the map $f : X_n/\!\!\sim \,\cong |\Gamma^n| \to \mathbb{R}^n$ by

$$x \mapsto \begin{pmatrix} \mu_1(A_{+1}(x)) - \mu_1(A_{-1}(x)) \\ \vdots \\ \mu_n(A_{+1}(x)) - \mu_n(A_{-1}(x)) \end{pmatrix}.$$

It is a continuous antipodal map, and hence has a zero that yields the desired partition.

Theorem 1.11. *Let $\mu_1, \ldots, \mu_n$ be n continuous probability measures on the unit interval. Then there exists a solution to the consensus $\frac{1}{2}$-division problem within the space $X_n/\!\!\sim \,\cong |\mathbb{Z}_2^{*(n+1)}| \cong |\Gamma^n|$. In particular, it suffices to make n cuts.* $\square$

Approximating a Solution

It is nice to know that a solution exists, but in practical applications how does one find (or at least approximate) a solution?

For this purpose let us call a division $I = A_{+1} \cup A_{-1}$ an *ε-approximate solution* if $\mu_i(A_{+1}), \mu_i(A_{-1}) \in [\frac{1}{2} - \varepsilon, \frac{1}{2} + \varepsilon]$ for all i. In order to find such a solution, consider an antipodally symmetric triangulation K subdividing Γ^n that is fine enough, i.e., for any edge xy in K, the inequality

$$|\mu_i(A_{+1}(x)) - \mu_i(A_{+1}(y))| \le \varepsilon$$

holds. This exists by continuity of the μ_i.

We will now construct an antipodal labeling $\lambda : \text{vert}(K) \to \{\pm 1, \ldots, \pm n\}$ of K. In terms of the introductory example with the two daughters, this labeling will tell us which family member considers the division most unfair and which of the two pieces he or she considers to be the smaller one. So, for a vertex x of K, let

$$\mu(x) = \min\{\mu_i(A_{+1}(x)), \mu_i(A_{-1}(x)) : i \in [n]\},$$

$$i(x) = \min\{i \in [n] : \mu_i(A_{+1}(x)) = \mu(x) \text{ or } \mu_i(A_{-1}(x)) = \mu(x)\}.$$

If $\mu(x) = \frac{1}{2}$, then x is an exact solution of the problem, and we are done. Otherwise, we define the labeling λ on x by

$$x \mapsto \begin{cases} +i(x), & \text{if } \mu_i(x)(A_{+1}(x)) = \mu(x), \\ -i(x), & \text{if } \mu_i(x)(A_{-1}(x)) = \mu(x). \end{cases}$$

If none of the vertices of K gave an exact solution, then λ is an antipodal labeling on K. By Tucker's lemma, there exist an edge xy in K and $i \in [n]$ such that $\lambda(x) = +i$ and $\lambda(y) = -i$. In particular, $\mu_i(A_{+1}(x)) < \frac{1}{2}$ and $\mu_i(A_{+1}(y)) > \frac{1}{2}$. Since the difference of $\mu_i(A_{+1}(x))$ and $\mu_i(A_{+1}(y))$ is at most ε, both of them will lie in the interval $[\frac{1}{2} - \varepsilon, \frac{1}{2} + \varepsilon]$. But by definition of the labeling, x turns out to be an ε-approximate solution.

As we have seen in the previous section, a complementary edge can be found by following a path in some graph. In particular, the ε-approximation to consensus $\frac{1}{2}$-division can be found using this procedure.

1.6 The Borsuk–Ulam Property for General Groups

In Sect. 1.3 we discussed the Borsuk–Ulam theorem and Tucker's lemma. One way to state the Borsuk–Ulam theorem is to say that every antipodal map $|(\mathbb{Z}_2)^{*(n+1)}| \cong |\Gamma^n| \cong \mathbb{S}^n \to \mathbb{R}^n$ has a zero. Tucker's lemma ensured the existence of a complementary edge in certain labelings of antipodally symmetric triangulations of $(\mathbb{Z}_2)^{*(n+1)} \cong \Gamma^n$.

Generalizing the Sphere

In this section we want to replace the group $\mathbb{Z}_2$ by an arbitrary finite group G and investigate analogous properties. The most important role, though, will be played by the cyclic groups $\mathbb{Z}_p$ of prime order.

The spheres that appeared as boundaries of the cross polytope in the previous section will now be replaced by more general spaces.

In the sequel we will make use of groups, group actions on a space, equivariant maps, connectivity of topological spaces, and shellability of simplicial complexes. A brief introduction to these concepts can be found in the Appendices D and B.

Definition 1.12. Let G be a finite group considered as a zero-dimensional geometric simplicial complex and let $N \geq 1$ be an integer. Let $E_N G$ be the geometric simplicial complex defined as the $(N + 1)$-fold join $E_N G = G * \cdots * G$ of G with itself. Its polyhedron is given by

$$|E_N G| = \left\{ (g_0 t_0, \ldots, g_N t_N) : g_i \in G, \, t_i \geq 0, \, \sum_{i=0}^{N} t_i = 1 \right\}.$$

The polyhedron $|E_N G|$ is a (compact) G-space via the diagonal action of G on all factors, i.e., for $g \in G$ and $(h_0 t_0, \ldots, h_N t_N) \in |E_N G|$ we define

$$g \cdot (h_0 t_0, \ldots, h_N t_N) = ((g h_0) t_0, \ldots, (g h_N) t_N).$$

To gain some better understanding of the geometric complex $E_N G$, consider the associated abstract simplicial complex. Its vertices may be identified with $(N + 1)$-dimensional vectors of the form $(\emptyset, \ldots, \emptyset, g, \emptyset, \ldots, \emptyset)$, where precisely one entry is a group element $g \in G$, and the faces may be identified with the set of vectors $(G \dot{\cup} \{\emptyset\})^{N+1}$. In other words, a face of $E_N G$ is determined by $N + 1$ choices of either the empty set or an element of G. The intersection of two faces given by vectors F and F' is then given by the vector $F \cap F'$ defined coordinatewise as follows:

$$(F \cap F')_i = \begin{cases} g, & \text{if } F_i = F_i' = g, \\ \emptyset, & \text{otherwise.} \end{cases}$$

The vector representation of the faces can be illustrated by arranging a grid of $k = |G|$ rows and $N + 1$ columns and for each face connecting any two subsequent grid elements that are not the empty set. Compare Fig. 1.19 demonstrating the case $k = 5$ and $N = 6$. The indicated face corresponds to the vector $(c, \emptyset, b, \emptyset, c, d, a)$, which in turn corresponds to all elements

$$\{(c t_0, 0, b t_2, 0, c t_4, d t_5, a t_6) : t_i \geq 0, \, t_0 + t_2 + t_4 + t_5 + t_6 = 1\}$$

of the geometric join.

Note that we can also read this diagram from top to bottom, and so we can identify $E_N G$ with the (polyhedron of a geometric realization of the) k-fold deleted join $(2^{\{0,\ldots,N\}})_{\Delta}^{*k}$ of the N-simplex $2^{\{0,\ldots,N\}}$. This is discussed in more detail on page 107 in Chap. 4.

A well-known example among these spaces is the boundary of the cross polytope, i.e., for $G = \mathbb{Z}_2$, the complex $E_N G$ can be identified with the boundary complex

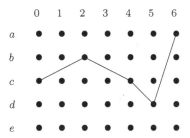

Fig. 1.19 A graphical representation of the faces of $E_N G$

Γ^N of the cross polytope. Indeed, an l-dimensional face of the cross polytope is given by a collection $\{\varepsilon_{i_0} e_{i_0}, \ldots, \varepsilon_{i_l} e_{i_l}\}$ with $i_0 < \cdots < i_l$ and $\varepsilon_{i_j} \in \{\pm 1\}$.

The following properties of the space $E_N G$ will play an essential role in the sequel.

Proposition 1.13. $|E_N G|$ *is a free G-space.*

Proof. Assume that

$$g \cdot (h_0 t_0, \ldots, h_N t_N) = (h_0 t_0, \ldots, h_N t_N).$$

Since there exists at least one j such that $t_j \neq 0$, we obtain $g h_j = h_j$, and hence g must be the neutral element in G.   □

Theorem 1.14. $E_N G$ *is a pure N-dimensional shellable simplicial complex, and hence has the homotopy type of a wedge of N-dimensional spheres. In particular, it is $(N - 1)$-connected.*

Proof. Let $G = \{g_1, \ldots, g_k\}$ be an arbitrary enumeration of the group elements. The maximal faces of $E_N G$ correspond to the set of vectors G^{N+1}. We order this set of vectors with respect to increasing lexicographic order, i.e.,

$$(g_{i_0}, \ldots, g_{i_N}) \prec (g_{j_0}, \ldots, g_{j_N})$$

if and only if there exists an $m \in \{0, \ldots, N\}$ such that $i_l = j_l$ for $l < m$ and $i_m < j_m$. For the case of $E_2 \mathbb{Z}_2$, the order in which the maximal faces appear is illustrated in Fig. 1.20, where $g_1 = +1$ and $g_2 = -1$.

We will now see that the defined order is indeed a shelling order. Let $F_1, \ldots, F_n$ be the ordered sequence of maximal faces of $E_N G$ and let $1 \leq j \leq n$. As stated in condition $(*)$ on page 178, we will show that $\left(\bigcup_{l=1}^{j-1} F_l \right) \cap F_j$ is a pure $(N - 1)$-dimensional complex. To see this, consider any intersection $F_i \cap F_j$ for $i < j$. We have to present an $i' < j$ such that $F_i \cap F_j \subseteq F_{i'} \cap F_j$ and the last intersection is $(N - 1)$-dimensional, i.e., $F_{i'}$ and F_j considered as vectors differ in exactly one coordinate. But this is easy. Let $F_i = (g_{i_0}, \ldots, g_{i_N})$ and $F_j = (g_{j_0}, \ldots, g_{j_N})$. By definition of the lexicographic order, there exists an m such that the first $m - 1$

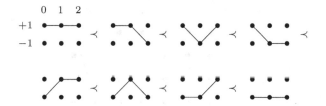

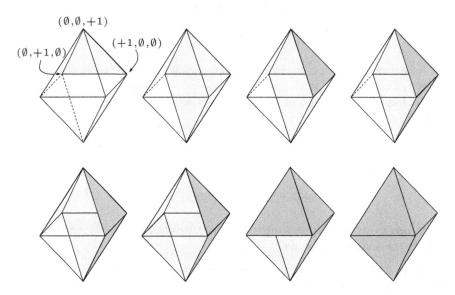

Fig. 1.20 The lexicographic shelling order in the case of $E_2 \mathbb{Z}_2$

coordinates of F_i and F_j are identical. Since $F_i \prec F_j$, we know that $i_m < j_m$. Consider

$$F = \left(g_{j_0}, \ldots, g_{j_{m-1}}, g_{i_m}, g_{j_{m+1}}, \ldots, g_{j_N} \right).$$

Then F is smaller than F_j, and hence there is an $i' < j$ such that $F_{i'} = F$. Certainly $F_{i'}$ has the desired properties. $\square$

A Generalization of the Borsuk–Ulam Theorem

We now approach the generalization of the Borsuk–Ulam theorem for the spaces $|E_N G|$. For this we need a replacement for the space $\mathbb{R}^n$ with the antipodal action. To this end, consider an N-dimensional real vector space $\mathbb{E}$ with a norm-preserving G-action, i.e., $\|gx\| = \|x\|$ for all $g \in G$, $x \in \mathbb{E}$. Furthermore, assume that the action is such that G does not fix any element except the origin, i.e., $\mathbb{E}^G = \{x \in \mathbb{E} : gx = x \text{ for all } g \in G\} = \{0\}$.

Definition 1.15 (Sarkaria [Sar00]). The group G has the *Borsuk–Ulam property* if for any $N \geq 1$ and any N-dimensional space $\mathbb{E}$ with norm-preserving G-action and $\mathbb{E}^G = \{0\}$, every continuous G-equivariant map $f : |E_N G| \to \mathbb{E}$ has a zero.

Note that the Borsuk–Ulam theorem guarantees that every continuous antipodal map $f : |E_N \mathbb{Z}_2| \to \mathbb{R}^N$ has a zero. We will now prove a far more general result for cyclic groups of prime order.

Theorem 1.16 (Bárány, Shlosman, Szűcs [BSS81] and Dold [Dol83]). *Let $G = \mathbb{Z}_p$ be the cyclic group of prime order $p \geq 2$. Then G has the Borsuk–Ulam property.*

Proof. Assume that there exist an N-dimensional space $\mathbb{E}$ with norm-preserving G-action such that $\mathbb{E}^G = \{0\}$ and a continuous G-equivariant map $f : |E_N G| \to \mathbb{E}$ without a zero.

The proof splits into considerations on the level of spaces and maps and purely algebraic considerations. We start with the spaces.

The map f induces a continuous map $\bar{f} : |E_N G| \to S(\mathbb{E})$ to the $(N-1)$-dimensional unit sphere $S(\mathbb{E})$ in $\mathbb{E}$ via $\bar{f}(x) = \frac{f(x)}{\|f(x)\|}$. Since

$$\bar{f}(gx) = \frac{f(gx)}{\|f(gx)\|} = \frac{gf(x)}{\|gf(x)\|} = \frac{gf(x)}{\|f(x)\|} = g\bar{f}(x),$$

the map $\bar{f}$ is G-equivariant. Note that G acts freely on $S(\mathbb{E})$ because p is prime and $\mathbb{E}^G = \{0\}$. Therefore, by Proposition D.13 on page 216, there exists a G-equivariant map $g : S(\mathbb{E}) \to |E_N G|$ since $|E_N G|$ is $(N-1)$-connected. Hence we have a composition

$$|E_N G| \xrightarrow{\bar{f}} S(\mathbb{E}) \xrightarrow{g} |E_N G|$$

of G-equivariant maps. By the equivariant simplicial approximation theorem, Theorem D.14, there exist a subdivision K of $E_N G$ and a simplicial map $\psi : K \to E_N G$ approximating $g \circ \bar{f}$.

We will now turn to the algebra and consider simplicial chain complexes and homology with coefficients in the field of rational numbers $\mathbb{Q}$.

First of all, note that $g \circ \bar{f}$ induces a map in homology

$$H_*(E_N G; \mathbb{Q}) \xrightarrow{H_*(\bar{f})} H_*(S(\mathbb{E}); \mathbb{Q}) \xrightarrow{H_*(g)} H_*(E_N G; \mathbb{Q}).$$

Secondly, ψ induces a map, $C_i(\psi) : C_*(K; \mathbb{Q}) \to C_*(E_N G; \mathbb{Q})$, of simplicial chain complexes. Note that there is a natural map, $C_*(E_N G; \mathbb{Q}) \to C_*(K; \mathbb{Q})$, that maps a generating i-simplex σ of $E_N G$ to the properly oriented sum of i-simplices of K contained in σ. In fact, the latter map induces the identity map in homology. Denote the composition of these two maps by $\bar{\psi} : C_*(K;\mathbb{Q}) \to C_*(K;\mathbb{Q})$.

For any i, consider the square-matrix representation of $\bar{\psi}_i : C_i(K;\mathbb{Q}) \to C_i(K;\mathbb{Q})$ with respect to the standard basis given by the (oriented) i-dimensional simplices of K. This matrix has only integral entries, since, so far, we could have used integral coefficients.

Now, since G acts freely and $\bar{\psi}_i$ is equivariant, each diagonal entry of the matrix appears p-fold, and hence the trace, $\mathrm{tr}(\bar{\psi}_i)$, is divisible by p. We now have the following:

$$\sum_{i=0}^{N}(-1)^i\,\mathrm{tr}(\bar{\psi}_i) = \sum_{i=0}^{N}(-1)^i\,\mathrm{tr}(H_i(\psi))$$

$$= \sum_{i=0}^{N}(-1)^i\,\mathrm{tr}(H_i(g \circ \bar{f}))$$

$$= \mathrm{tr}\left(H_0(g \circ \bar{f})\right) + (-1)^N\,\mathrm{tr}\left(H_N(g \circ \bar{f})\right)$$

$$= \mathrm{tr}(H_0(g \circ \bar{f}))$$

$$= 1.$$

The first equality is given by the Hopf trace formula, as discussed in Theorem B.72 on page 193, and the fact that ψ and $\bar{\psi}$ induce the same map in homology; the second holds since $\psi \simeq g \circ \bar{f}$ are homotopic maps; the third equation follows from the fact that $|E_N G|$ is a wedge of N-spheres; the fourth is due to the fact that $H_N(g \circ \bar{f})$ factors through $H_N(S(\mathbb{E});\mathbb{Q}) = 0$; and finally, the last equality follows from the fact that $g \circ \bar{f}$ is a nontrivial map between two nonempty spaces. We have reached a contradiction, since 1 is not divisible by p.   □

The following theorem is a generalization of Theorem 1.16 to the case of powers of cyclic groups of prime order. The proof of this generalization requires some methods that go beyond those of this book.

Theorem 1.17 (Özaydin [Öza87], Volovikov [Vol96], and Sarkaria [Sar00]). *Let $p \geq 2$ be a prime and $r \geq 1$. The group $G = (\mathbb{Z}_p)^r$ has the Borsuk–Ulam property.*   □

1.7   Consensus $\frac{1}{k}$-Division

In analogy to Sect. 1.5, we are faced with the problem of dividing a piece of inherited land among k siblings—instead of two—such that all n family members believe that all siblings receive a piece of land of the same value. At this point we also want to introduce another common interpretation of the situation due to Noga Alon [Alo87].

Fig. 1.21 A necklace with beads of several types

Assume that k thieves have stolen necklace like the necklace in Fig. 1.21 with n different types of precious stones such that the number of times each type occurs is divisible by k. The thieves are about to divide the necklace among them such that:

- They do as few cuts as possible,
- Each gets the same number of beads of each type.

Assuming that the necklace has been nonviolently opened at the clasp, what is the fewest number of cuts that always suffices?

If one generalizes and considers each type of bead to be continuously distributed along the necklace, we arrive at the generalization of the problem in Sect. 1.5.

Let $n \geq 1, k \geq 2$, and $\mu_1, \ldots, \mu_n$ be continuous probability measures on the unit interval $I = [0, 1]$. In this more general situation, we are now looking for divisions $I = A_1 \cup \cdots \cup A_k$ such that $\mu_i(A_j) = \frac{1}{k}$ for all $i = 1, \ldots, n$ and $j = 1, \ldots, k$. Note again that these conditions imply $\mu_i(A_j \cap A_{j'}) = 0$ for all i and $j \neq j'$.

In order to model the situation nicely, we need to determine how many cuts we need at least. Just imagine a necklace with $n \cdot k$ beads arranged in such a way that each type comes in a block of k. Then we will certainly need to make $k - 1$ cuts within each block and hence need at least $n(k - 1)$ cuts in total. Clearly, in the continuous situation we will not be able to deal with fewer cuts either. In Fig. 1.22, a possible situation is shown for $n = 3, k = 4$. The measures μ_i are given by their respective density functions φ_i.

$E_N G$ as Solution Space

Experienced by now, we easily come up with a good space to model our situation. The space needs to encode $N = n(k - 1)$ cuts, and for each of the $N + 1$ resulting

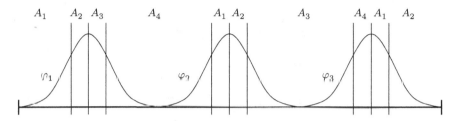

A_1 A_2 A_3 A_4 A_1 A_2 A_3 A_4 A_1 A_2

Fig. 1.22 At least $n(k-1)$ cuts are necessary

intervals, there has to be an assignment into which of the sets A_i, $i = 1,\ldots,k$, it belongs:

$$X_{N,k} = \left\{ \big((g_0,t_0),\ldots,(g_N,t_N)\big) : g_i \in [k], t_i \ge 0, \sum_{i=0}^{N} t_i = 1 \right\}.$$

As before, t_i is encoding the length of the $(i+1)$st interval, and g_i defines that this interval belongs to A_{g_i}. Moreover, we introduce the same equivalence relation:

$$\big((g_0,t_0),\ldots,(g_j,t_j),\ldots,(g_N,t_N)\big) \sim \big((g_0,t_0),\ldots,(g_j',t_j),\ldots,(g_N,t_N)\big)$$

if and only if for all j, either $g_j = g_j'$ or $t_j = 0$. By the definition of a join, we see that $X_{N,k}/\!\sim$ can be identified with $|\tilde{E}_N G|$, where G is any group of order k.

Theorem 1.18 (Continuous necklace theorem [Alo87]). *Let $n \ge 1, k \ge 2$ and $\mu_1,\ldots,\mu_n$ be continuous probability measures on the unit interval. Then there exist $n(k-1)$ cuts of the interval and an assignment of the resulting intervals to sets $A_1,\ldots,A_k$ such that $\mu_i(A_j) = \frac{1}{k}$ for all i,j.*

Since we know that the Borsuk–Ulam property holds for the cyclic groups $G = \mathbb{Z}_p$ of prime order, we will first reduce the necklace problem to the case $k = p$, $p \ge 2$ prime.

Lemma 1.19 (Reduction to prime case). *Let $n \ge 1, k_1, k_2 \ge 2$. Assume that the necklace theorem is true for the parameters n and k_1 and n and k_2. Then it is also true for the parameters n and $k_1 k_2$.*

Proof. Let $\mu_1,\ldots,\mu_n$ be given, and set $N_1 = n(k_1 - 1)$, $N_2 = n(k_2 - 1)$, and $N = n(k_1 k_2 - 1)$. By assumption, there exist N_1 cuts of the interval I and an assignment of the resulting intervals to sets $A_1,\ldots,A_{k_1}$ such that $\mu_i(A_j) = \frac{1}{k_1}$ for all i, j. Now, for some fixed j, consider the set A_j that is a union of intervals. By restricting the μ_i to A_j, rescaling them by a factor k_1, and gluing the intervals together, we obtain a new instance of the problem. Hence, by assumption, for all $j = 1,\ldots,k_1$, there exist N_2 cuts within A_j and an assignment of the resulting intervals to sets $A_1^j, \ldots, A_{k_2}^j$ such that $\mu_i(A_l^j) = \frac{1}{k_1 k_2}$ for all i, j, l. Therefore, in total we have

$$N_1 + k_1 N_2 = n(k_1 - 1) + k_1 n(k_2 - 1) = N$$

cuts of the interval, and the sets $A_l^j, , j = 1, \ldots, k_1, l = 1, \ldots, k_2$ are the desired sets.   □

Proof (of the necklace theorem). By the previous lemma we can assume $k = p$ to be a prime number $p \geq 2$. Let $G = \mathbb{Z}_p$ be the cyclic group of order p and $N = n(p - 1)$. The space $|E_N G|$ encodes the possible solutions. Let $\mathbb{E}$ be the set of all real $n \times p$ matrices with the property that the elements of each row sum to zero. In particular, for any such matrix, a row can have all entries equal only if it is a zero row. The group G acts on $\mathbb{E}$ by cyclically permuting columns. The action is obviously norm-preserving, $\mathbb{E}^G = \{0\}$ by the previous observation, and $\dim \mathbb{E} = n(p - 1) = N$. We define a map $f : |E_N G| \to \mathbb{E}$ as follows. Each $x \in |E_N G|$ gives rise to a subdivision $I = A_1(x) \cup \cdots \cup A_p(x)$ continuously depending on x as described at the beginning of this section. The group acts on $|E_N G|$ in such a way that the resulting sets, $A_1(x), \ldots, A_p(x)$, are permuted cyclically. Now define the matrix $f(x)$ entrywise by

$$\left(f(x) \right)_{i,j} = \left(\mu_i(A_j(x)) - \frac{1}{p} \right).$$

Then f is well defined and yields a continuous G-equivariant map. Hence, by Theorem 1.16, f has a zero. But a zero of f corresponds exactly to a desired fair necklace splitting.   □

Exercises

1. Show that in the algebraic proof of Sperner's lemma on page 5, the induction hypothesis indeed yields $\lambda_{n-1}(c_i) = \tau_i, i = 1, \ldots, n + 1$.
2. Show that in the derivation of Sperner's lemma with Brouwer's fixed-point theorem on page 7, the map $f : |\Delta^n| \to |\Delta^n|$ has no fixed points on the boundary.
3. This exercise is about deriving the *fundamental theorem of algebra* directly from Brouwer's fixed-point theorem. The fundamental theorem of algebra states that any nonconstant polynomial $p = z^n + a_{n-1}z^{n-1} + \cdots + a_1 z + a_0$ with complex coefficients has a zero in the field of complex numbers.
   In order to prove this, let $R = 2 + |a_0| + \cdots + |a_{n-1}|$, and define

   $$f : \{z \in \mathbb{C} : |z| \leq R\} \longrightarrow \{z \in \mathbb{C} : |z| \leq R\}$$

$$z = r\,e^{i\varphi} \longmapsto \begin{cases} z - \frac{p(z)}{R}, & \text{if } r = 0, \\ z - \frac{p(z)}{R\,e^{i\varphi r(n-1)}}, & \text{if } 0 < r \le 1, \\ z - \frac{p(z)}{R z^{n-1}}, & \text{if } 1 \le r. \end{cases}$$

Show that f is a well-defined continuous map and deduce the fundamental theorem of algebra. This proof is due to B.H. Arnold [Arn49].

4. Show that it is always possible to label the vertices of $\mathrm{sd}^k \Delta^{n-1}$ with labels $p_1, \ldots, p_n$ such that each $(n-1)$-simplex obtains all labels. In other words, show that the 1-skeleton of $\mathrm{sd}^k \Delta^{n-1}$ has chromatic number n. This is needed in the proof for the existence of envy-free fair division solutions on page 10.

5. Provide the details that show that the division $(A_1, \ldots, A_n; \pi)$ defined on page 11 in the proof of Theorem 1.5 is indeed envy-free.

6. Solve the following *rental harmony problem* by Francis E. Su [Su99].

   Suppose n students want to share an apartment with n rooms that they have rented for some fixed price. Now they are to decide who gets which room and for what part of the total rent. Moreover, assume that the following three conditions are satisfied:

   (a) **(Good house)** In any partition of the rent, each person finds some room acceptable.

   (b) **(Miserly tenants)** Each person always prefers a free room to a room for which they have to pay.

   (c) **(Closed preference set)** If a person is satisfied with a certain room for a convergent sequence of prices, then he is also satisfied with this room at the limit price.

   Give a solution that maintains harmony among the students, i.e., a solution in which no one would like to switch rooms.

7. In this exercise we want to derive the classical *ham sandwich theorem* from the Borsuk–Ulam theorem. Consider a sandwich consisting of two slices of bread and one slice of ham. Show that it can be cut with one straight cut in such a way that each of the three pieces is divided in half. More generally, show the following. Given n continuous probability measures in $\mathbb{R}^n$, show that there exists an affine hyperplane such that each measure takes the value $\frac{1}{2}$ on each of the two half-spaces defined by the hyperplane. In order to do so, find a natural way to describe a pair of half spaces defined by an affine hyperplane in $\mathbb{R}^n$ with the help of the n-sphere $\mathbb{S}^n \subseteq \mathbb{R}^{n+1}$.

8. Deduce Tucker's lemma from the Borsuk–Ulam theorem as stated on page 15.

9. Show that Tucker's lemma in fact holds for arbitrary antipodally symmetric triangulations of the n-dimensional sphere, i.e., the simplicial complex K in the formulation of the lemma does not have to be a subdivision of Γ^n.

10. Show that in the definition of an ε-almost-alternating simplex on page 16, the sign ε is well defined.

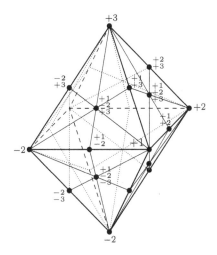

Fig. 1.23 The first barycentric subdivision of Γ^2

11. Show that as required at the end of the proof of the Ky Fan theorem (weak version) on page 19, the vertices $\{\pm e_1\}$ are not connected by a path in G.

12. Freund und Todd gave a combinatorial proof of Tucker's lemma [FT81]. In this exercise we want to follow their ideas in order to obtain an appealing proof of Tucker's lemma in a very special situation. We will later use this to obtain a combinatorial proof of Kneser's conjecture. Let $K = \mathrm{sd}^1 \Gamma^{n-1}$ be the first barycentric subdivision of the boundary complex of the cross polytope and let $\lambda : \mathrm{vert}(K) \to \{\pm 1, \ldots, \pm(n-1)\}$ be an antipodal labeling. We assume, contrary to Tucker's lemma, that there does not exist a complementary edge. Identify the vertices $\{\pm e_1 \ldots, \ldots \pm e_n\}$ of Γ^{n-1} with the set $\{\pm 1, \ldots, \pm n\}$ via $\pm e_i \leftrightarrow \pm i$. A vertex of K is then given by a nonempty subset $v \subset \{\pm 1, \ldots, \pm n\}$ with $v \cap -v = \emptyset$. Moreover, a simplex of K is given by an inclusion chain of such sets $\sigma = \{v_0 \subset \cdots \subset v_k\}$. For an illustration, see Fig. 1.23.

    We call a simplex σ *complete* if $\lambda(\sigma) = \{\lambda(v_0), \ldots, \lambda(v_k)\} \supseteq (v_k \setminus \{\pm n\})$. We define a graph G whose vertex set is the set of all complete simplices, and for two complete simplices $\tau \subset \sigma = \{v_0 \subset \cdots \subset v_k\}$, τ is adjacent to σ if and only if $\lambda(\tau) \supseteq (v_k \setminus \{\pm n\})$. In order to reach a contradiction, show the following. The simplices $\sigma_\pm = \{\{\pm n\}\}$ are complete and of degree 1 in G, all other vertices of G have degree 2, but $\sigma_\pm$ are not the endpoints of a path in G.

    Note that the last vertex v_k in $\sigma = \{v_0 \subset \cdots \subset v_k\}$ completely describes which coordinate orthant contains σ. This information was used by Freund and Todd to obtain a proof of Tucker's lemma in the general case by the same construction of the graph. They essentially replaced $v_k \setminus \{\pm n\}$ by the coordinates of the hyperorthants, disregarding the nth direction.

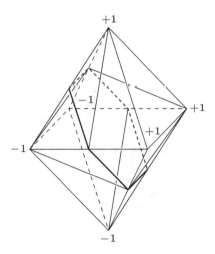

Fig. 1.24 An invariant closed loop

13. Generalize the Ky Fan theorem (strong version) to n-pseudomanifolds. An n-*dimensional pseudomanifold* K is a simplicial complex in which all maximal simplices are of dimension n, and any $(n-1)$-dimensional simplex is contained in at most two n-dimensional simplices. The *boundary of K*, denoted by ∂K, is the complex given by the $(n-1)$-dimensional simplices of K that are contained in exactly one n-dimensional simplex, together with all of their faces.

14. Let K be an antipodally symmetric triangulation of the 2-dimensional sphere, and let $\lambda : \text{vert}(K) \to \{\pm 1\}$ be an equivariant labeling, i.e., $\lambda(-v) = -\lambda(v)$. The labeling λ induces a continuous map $\bar{\lambda} : |K| \to [-1, +1]$ to the interval as shown in Exercise 29 on page 197. Show that there exists a closed loop c in $|K|$ that is invariant under the antipodal action, i.e., $c = -c$, and that is mapped to zero under $\bar{\lambda}$. A very easy example is shown in Fig. 1.24.

15. This exercise is concerned with an alternative proof that $|E_N G|$ is (up to homotopy) a wedge of spheres. Consider three compact pairwise disjoint spaces X, Y, Z, and $x_0 \in X$, $y_0 \in Y$. Show that there is a homotopy equivalence

$$(X \vee Y) * Z \simeq (X * Z) \vee (Y * Z),$$

    or more precisely

$$((X \cup Y)/x_0 \sim y_0) * Z \simeq (X * Z \cup Y * Z)/(x_0, 0) \sim (y_0, 0).$$

    In other words, the wedge and join operations of spaces respect a distributivity law.

16. Use the previous exercise to prove that $|E_N G|$ is a wedge of N-dimensional spheres. Determine the number of spheres involved. Hint: Realize the zero dimensional geometric complex G as a wedge of 0-spheres.

17. Show that in the proof of Theorem 1.16, $\text{tr}(\bar{\psi})$ is indeed divisible by p.
18. Modify the proof of Theorem 1.16 in order to prove the following. Let $G = \mathbb{Z}_p$, where $p \geq 2$ is prime, let $\mathbb{E}$ be an N-dimensional real vector space with a linear G-action, and let $\mathbb{E}^G = \{0\}$. Then every continuous G-equivariant map $f : |E_N G| \to \mathbb{E}$ has a zero.
19. Modify the proof of Theorem 1.16 in order to prove the following. Let $G = \mathbb{Z}_p$, where $p \geq 2$ is prime, and $n \geq 1$. Then there is no G-equivariant map $f : |E_n G| \to |E_{n-1} G|$.
20. Show that any finite group G contains $\mathbb{Z}_p$ as a subgroup for some prime $p \geq 2$.
21. Let G be any finite group. Use the previous two exercises in order to prove that there is no G-equivariant map $f : |E_n G| \to |E_{n-1} G|$. Hint: Proposition D.13.
22. This exercise proves a theorem by Dold [Dol83]. Let X and Y be G-spaces such that Y is a free G-space. Assume that there exists a G-equivariant map $f : X \to Y$. Then $\text{conn } X \leq \dim Y - 1$. Hint: Proposition D.13 and the previous exercise.
23. Show that the divisions $I = A_1 \cup \cdots \cup A_k$ considered on page 29 have the property that $\mu_i(A_j \cap A_{j'}) = 0$ for all i, j, j' with $j \neq j'$. Hence the divisions of the interval are partitions of the interval with respect to the measures.
24. Solve the discrete necklace problem, which is the following. Let $n, k \geq 2$, and let $m_1, \ldots, m_n \geq 2$ be any set of numbers, each divisible by k. Consider an open necklace consisting of $m_1 + \cdots + m_n$ beads of n different types such that among them, m_i are of type i for each i. Then there exist $n(k-1)$ cuts and a division of the resulting pieces among k thieves such that each thief obtains $\frac{m_i}{k}$ beads of type i.

Chapter 2
Graph-Coloring Problems

A very important graph parameter is the chromatic number. For a given graph, it is the smallest number of colors for which a coloring of the vertices exists such that adjacent vertices receive different colors. The search for a proof of the four color theorem—stating that every planar map can be colored with four colors such that adjacent countries receive different colors (Fig. 2.1)—has certainly been one of the driving sources [Ore67, Saa72, Tho98] of graph theory for a long time. Presently, graph coloring plays an important role in several real-world applications and still engages exciting research.

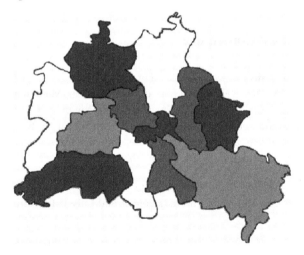

Fig. 2.1 A proper four-coloring of a map

In this chapter we will present an important side story, the story of a conjecture formulated by Martin Kneser in 1955 that remained unsolved until 1977 (published in 1978 [Lov78]). The revolutionary method with which László Lovász settled the notorious conjecture can be seen as the origin of the field with which this book deals. Guided by some deep insight, Lovász associated a simplicial complex to a

M. de Longueville, *A Course in Topological Combinatorics*, Universitext,
DOI 10.1007/978-1-4419-7910-0_2,
© Springer Science+Business Media New York 2013

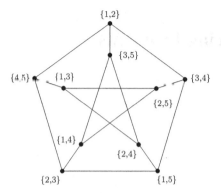

Fig. 2.2 The Petersen graph in the guise of the Kneser graph $KG_{5,2}$

graph in such a way that the topology of the complex provides some information about the chromatic number of the graph, thereby transforming a discrete problem into a topological one. The main tool he employed was the Borsuk–Ulam theorem. His proof, and the efforts to understand it, have triggered a considerable amount of research. By now, Lovász's original proof has gone through many transformations and inspired alternative proofs even until very recently. We will touch upon most of the ideas involved in the several proofs that emerged over the last decades.

2.1 The Kneser Conjecture

Kneser's original conjecture was published in 1955 as an exercise in *Jahresberichte der DMV* [Kne55], the yearly account of the German Mathematical Society. Apparently, at this time Kneser has been thinking about quadratic forms. But the connection to the conjecture seems to have been forgotten. Kneser stated his conjecture originally as a problem about sets, but the translation into a graph-theoretic problem is straightforward. Denote the set of k-subsets of $[n]$ by $\binom{[n]}{k}$.

Definition 2.1. The *Kneser graph* $KG_{n,k}$ for $n \geq 2$, $k \geq 1$, has vertex set $\binom{[n]}{k}$, and any two vertices $u, v \in \binom{[n]}{k}$ are adjacent if and only if they are disjoint, i.e., $u \cap v = \emptyset$.

To get a feeling for these graphs, let's consider first some cases with easy parameters. For $k = 1$, we obtain the complete graph on n vertices, i.e., any two vertices are adjacent. For $n = 2k$, each k-set is adjacent to and only to its complement. In other words, we obtain a complete matching, i.e., a set of disjoint edges covering all vertices. For $2k > n$, we obtain a set of vertices without any edges. For this reason, we will restrict ourselves to the cases $2k \leq n$ in the sequel. The first interesting case appears already for $n = 5$ and $k = 2$. Figure 2.2 shows the Kneser graph $KG_{5,2}$ which is the famous Petersen graph [Wes05, Die06].

As was already observed by Kneser, $KG_{n,k}$ admits a proper coloring with $n - 2k + 2$ colors simply as follows:

$$c : \binom{[n]}{k} \longrightarrow [n - 2k + 2],$$

$$u \longmapsto \min\{\min\{x : x \in u\}, n - 2k + 2\}.$$

We have to check that vertices receiving the same color are not adjacent. Consider two vertices $u, v \in \binom{[n]}{k}$ with $c(u) = c(v) = c$. If $c < n - 2k + 2$, then $c \in u \cap v$; otherwise, $u, v \subseteq \{n - 2k + 2, \ldots, n\}$. But $\{n - 2k + 2, \ldots, n\}$ contains $2k - 1$ elements, and hence u and v cannot be disjoint. The coloring witnesses the upper bound $\chi(KG_{n,k}) \leq n - 2k + 2$ for the chromatic number of the Kneser graph. Kneser conjectured that this bound was sharp, in other words, it is not smaller than $n - 2k + 2$.

Theorem 2.2 (Lovász [Lov78]). *The chromatic number of the Kneser graph $KG_{n,k}$ is $n - 2k + 2$.*

We will discuss Lovász's proof in more detail in the next section. After Imre Bárány had learned about Lovász' proof in 1978, he came up with a fairly short proof of Kneser's conjecture. Both proofs have different strengths. While Lovász's proof involves a theorem of deep insight that yields a lower bound for the chromatic number of any graph, and then specializes to the family of Kneser graphs, Bárány's proof is a fairly direct and elegant application of the Borsuk–Ulam theorem, but does not shed as much light on general graph-coloring problems.

The first proof we will discuss is the most recent proof by Greene [Gre02]. It is a tricky simplification of Bárány's proof.

Proof (topological). Assume that for some n and k the chromatic number of $KG_{n,k}$ is less than $n - 2k + 2$, and let $c : \binom{[n]}{k} \to \{1, \ldots, n - 2k + 1\}$ be a proper coloring. Set $d = n - 2k + 1$ and choose a set X of n vectors on the d-dimensional sphere $\mathbb{S}^d$ such that any $d + 1$ of them constitute a linearly independent set. Identify these n vectors with the ground set $[n]$. In other words, each vertex of $KG_{n,k}$ corresponds to a set of k vectors on the sphere. In order to apply the Borsuk–Ulam theorem, we will construct d open sets $U_1, \ldots, U_d$ and one closed set A covering $\mathbb{S}^d$. Let

$$U_i = \left\{x \in \mathbb{S}^d : \text{there exists a } k\text{-set } S \subset X, c(S) = i, S \subset H(x)\right\},$$

where $H(x) = \{y \in \mathbb{S}^d : \langle x, y \rangle > 0\}$ is the open hemisphere with pole x.

Now let $A = \mathbb{S}^d \setminus (U_1 \cup \cdots \cup U_d)$ be the complement. We show that none of the sets contains a pair of antipodal points, hence obtaining a contradiction to the Borsuk–Ulam theorem, Theorem 1.6(4).

Consider $x \in U_i$, i.e., $H(x)$ contains a k-subset of X colored with color i. Since $H(x)$ and $H(-x)$ are disjoint and c is a proper coloring, $H(-x)$ cannot contain a k-subset of X colored with i as well, and hence $-x \notin U_i$.

Now assume $\pm x \in A$. By definition of A, neither $H(x)$ nor $H(-x)$ contains a k-subset of X. Hence there must be at least $n - 2(k - 1) = n - 2k + 2 = d + 1$ points of X lying on the equator $\{y \in \mathbb{S}^d : \langle x, y \rangle = 0\}$, which is contained in a subspace of dimension d. This contradicts the condition that any $d + 1$ vectors of X are linearly independent.   □

The second proof we discuss is quite recent as well and due to Jiří Matoušek [Mat04]. It is considered to be the first combinatorial proof of Kneser's conjecture. The topological chore is an application of Tucker's lemma in the very special case that the vertices of the first barycentric subdivision of ∂Q^{n+1} are labeled. Exercise 12 on page 33 is concerned with a simple combinatorial proof for Tucker's lemma in this case.

Proof (combinatorial). Assume that there is a proper coloring $c : \binom{[n]}{k} \to \{2k - 1, 2k, \ldots, n - 1\}$ of $KG_{n,k}$ with $n - 2k + 1$ colors. This will eventually yield a contradiction to Tucker's lemma. For this we need a little notation. Let $K = \mathrm{sd}^1 \, \partial Q^n$ be the first barycentric subdivision of the boundary of the cross polytope Q^n. As explained in Exercise 12 on page 33, we can identify $\mathrm{vert}(K)$ with the set $\mathcal{Q}_n$ of nonempty subsets $v \subset \{\pm 1, \ldots, \pm n\}$ such that $v \cap -v = \emptyset$. For any $S \in \mathcal{Q}_n$, let $S_+ = \{i : i > 0, i \in S\}$, resp. $S_- = \{i : i > 0, -i \in S\}$. Let $\succeq$ be an arbitrary linear order of the subsets of $[n]$ such that whenever $|A| > |B|$, then $A \succ B$. The existence of such an extension is the subject of Exercise 1 in the appendix on page 206. Now for any $A \subseteq [n]$ with $|A| \geq k$, set $\bar{c}(A) = c(A')$, where $A' \subseteq A$ is the set of the k smallest numbers in A.

We now define a labeling $\lambda : \mathcal{Q}_n \to \{\pm 1, \ldots, \pm(n-1)\}$ by

$$
\lambda(S) = \begin{cases}
|S|, & \text{if } |S| < 2k - 1 \text{ and } S_+ \succeq S_-, \\
-|S|, & \text{if } |S| < 2k - 1 \text{ and } S_- \succeq S_+, \\
\bar{c}(S_+), & \text{if } |S| \geq 2k - 1 \text{ and } S_+ \succeq S_-, \\
-\bar{c}(S_-), & \text{if } |S| \geq 2k - 1 \text{ and } S_- \succeq S_+.
\end{cases}
$$

This labeling is antipodal and does not yield complementary edges and therefore yields the desired contradiction. The antipodality is obvious, and in order to see that there are no complementary edges, assume to the contrary that there are $S, T \in \mathcal{Q}_n$ forming an edge in K with $\lambda(S) = -\lambda(T) > 0$. By definition of barycentric subdivision, S is contained in T or vice versa. Since the other case works analogously, we consider only the case $S \subset T$. Hence $|S| < |T|$, and by definition of λ, we must have $\bar{c}(S_+) = \bar{c}(T_-)$. But $S_+ \cap T_- \subseteq T_+ \cap T_- = \emptyset$, which yields a contradiction to the fact that c was a proper coloring.   □

2.2 Lovász's Complexes

In this section we will associate several simplicial complexes to graphs. All of these constructions are due to László Lovász [Lov78, BK07]. These complexes will be used to give lower bounds on the chromatic number.

While the conditions on a proper coloring of a graph—no monochromatic edge— is a local condition, the chromatic number captures a global phenomenon. A good example is an odd cycle, which has chromatic number 3. It can be colored vertex for vertex along the cycle with two colors until the last vertex, where the true value of the chromatic number is revealed. In order to obtain bounds for the chromatic number, it is therefore necessary to capture the global behavior of the graph in some way. There are quite a few global invariants for topological spaces. In that respect, it seems natural to try to assign a topological space to a graph in such a way that the global topological properties of the space reflect some global property of the graph.

The Neighborhood Complex

We will now describe Lovász's neighborhood complex, the first construction of a simplicial complex that we associate with a graph.

Let $G = (V, E)$ be a finite simple graph. Let the *neighborhood complex* $\mathcal{N}(G)$ be the simplicial complex with vertex set V and simplices given by subsets $A \subseteq V$ such that all vertices in A have a common neighbor. As a first example consider Fig. 2.3.

Note that the neighborhood complex of a graph without edges is empty, and as soon as the graph has an edge it is nonempty.

The neighborhood complex of an odd cycle is an odd cycle of the same length. In fact, if the odd cycle has the vertex set $\{0, 1, \ldots, 2k\}$ in such a way that two vertices

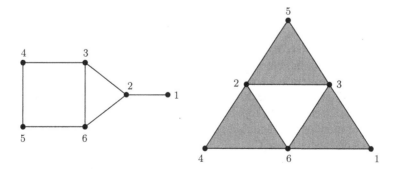

Fig. 2.3 A graph G along with its neighborhood complex

Fig. 2.4 Neighborhood complex of an odd cycle, of an even cycle, and of a bipartite graph

are adjacent if and only if they differ by one modulo $2k + 1$, then the neighborhood complex is a 1-dimensional complex with edge set

$$\{\{0, 2\}, \{2, 4\}, \ldots, \{2k - 2, 2k\}, \{2k, 1\}, \{1, 3\}, \ldots, \{2k - 1, 0\}\}.$$

In the same manner, the neighborhood complex of an even cycle (of length greater than or equal to 6) consists of two cycles, each half the length of the original cycle. And more generally, if $G = (V \dot\cup V', E)$ is a bipartite graph with independent sets V and V', then each simplex of $\mathcal{N}(G)$ is contained in either V or V', and hence the complex is not connected. Figure 2.4 illustrates all of these cases, where in each case the graph is sketched in black and the neighborhood complex in gray. In the case of the third bipartite graph, the facets of the neighborhood complex consist of a 3-dimensional simplex and two 2-dimensional simplices sharing an edge.

As a last class of examples, we consider the neighborhood complexes $\mathcal{N}(K_n)$ of the complete graph K_n. Let us denote the vertex set of K_n by $[n]$. Then each nonempty proper subset $A \subset [n]$ has a common neighbor and therefore is a simplex of $\mathcal{N}(K_n)$. Thus, $\mathcal{N}(K_n)$ is the boundary complex of the simplex on the vertex set $[n]$ and hence a sphere of dimension $n - 2$.

We have already found some indication of the phenomenon that global properties of the neighborhood complex capture information about the chromatic number. If the neighborhood complex of a graph is nonempty, then the graph has at least one edge and therefore has chromatic number at least two. If we encounter a nonempty connected neighborhood complex of a graph G, then we already know that it cannot be bipartite and hence has chromatic number at least three.

The emerging pattern is perpetuating, as the following theorem says. We will provide an easy proof on page 50. Before we state Lovász's theorem, we should briefly remind ourselves of the topological notion of k-connectedness as defined on page 170. For more on this, and the subsequently used concepts of order topology, we refer to Appendices B and C.

Theorem 2.3 (Lovász [Lov78]). *Let $G = (V, E)$ be a finite simple graph. If the neighborhood complex $\mathcal{N}(G)$ of G is k-connected, then the graph has chromatic number at least $k + 3$. In other words,*

$$\chi(G) \geq \operatorname{conn}(|\mathcal{N}(G)|) + 3.$$

Observe the general applicability of the theorem. Whenever we are interested in the chromatic number of a graph G, we may determine its neighborhood complex, and every lower bound we obtain for the connectivity of $\mathcal{N}(G)$ yields a lower bound for the chromatic number of G. In particular, Lovász obtained the first proof of the Kneser conjecture by showing that $\mathrm{conn}\,(|\mathcal{N}(KG_{n,k})|) = n - 2k - 1$. We start by giving a very beautiful and short proof of this fact.

Proposition 2.4. *The neighborhood complex $\mathcal{N}(KG_{n,k})$ of the Kneser graph is homotopy equivalent to a wedge of spheres of dimension $n - 2k$. In particular,* $\mathrm{conn}\,(|\mathcal{N}(KG_{n,k})|) = n - 2k - 1$.

Proof. We prove that the face poset $\mathcal{F}(\mathcal{N}(KG_{n,k}))$ of $\mathcal{N}(KG_{n,k})$, i.e., the set of nonempty faces of $\mathcal{N}(KG_{n,k})$ partially ordered by inclusion, is homotopy equivalent to the following part, $B_{n,k}$, of the Boolean lattice

$$B_{n,k} = \{S \subseteq [n] : k \leq |S| \leq n - k\}$$

By Corollary C.10 on page 206, the poset $B_{n,k}$ is lexicographically shellable. In particular, it has the homotopy type of a wedge of spheres of dimension $n - 2k$.

Define the order-preserving maps

$$f : \mathcal{F}(\mathcal{N}(KG_{n,k})) \longrightarrow B_{n,k},$$

$$F \longmapsto \bigcup F = \{x \in [n] : \text{exists } v \in F \text{ such that } x \in v\},$$

and

$$g : B_{n,k} \longrightarrow \mathcal{F}(\mathcal{N}(KG_{n,k})),$$

$$A \longmapsto \binom{A}{k} = \{S \subset A : |S| = k\}.$$

Then obviously $f \circ g = \mathrm{id}_{B_{n,k}}$, and for each $F \in \mathcal{F}(\mathcal{N}(KG_{n,k}))$, we have $\mathrm{id}_{\mathcal{F}(\mathcal{N}(KG_{n,k}))}(F) = F \subseteq (g \circ f)(F)$. By the order homotopy lemma, Lemma C.3, the maps $\mathrm{id}_{\mathcal{F}(\mathcal{N}(KG_{n,k}))}$ and $g \circ f$ are homotopic, and hence f is a homotopy equivalence. $\square$

The Neighbor Set Function v

Returning to the situation of a general graph G, we define v to be the *neighbor set function*, i.e., for a subset $A \subseteq V$ of vertices of G, define $v(A)$ to be the set of all vertices in V that are adjacent to all vertices in A, i.e.,

$$v(A) = \{v \in V : v \text{ is adjacent to } a \text{ for all } a \in A\}.$$

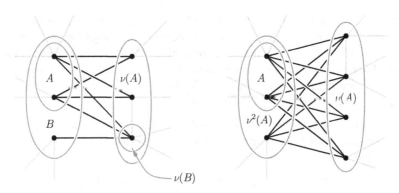

Fig. 2.5 Proof of statements 1 and 2 in Proposition 2.5

Note that for each $A \subseteq V$, the two sets A and $\nu(A)$ are the vertex sets of a complete bipartite subgraph of G, i.e., each vertex of A is adjacent to each vertex of $\nu(A)$.

With this notation, the simplices of $\mathcal{N}(G)$ are precisely given by subsets $A \subseteq V$ such that $\nu(A) \neq \emptyset$.

Proposition 2.5. *Let* $G = (V, E)$ *be a finite simple graph. The neighbor set function* $\nu : \mathcal{P}(V) \to \mathcal{P}(V)$ *of the power set of V to itself satisfies the following:*

1. If $A \subseteq B \subseteq V$, *then* $\nu(B) \subseteq \nu(A)$.
2. $A \subseteq \nu^2(A)$ *for any* $A \subseteq V$.
3. $\nu(A) = \nu^3(A)$ *for any* $A \subseteq V$.

Proof. For the first two assertions we confine ourselves to a "proof by picture" [Pól56] as given in Fig. 2.5.

The third statement is an obvious application of the first two.   □

Note that the function $g \circ f$ in the proof of Proposition 2.4 is the function ν^2 for the Kneser graph. In detail,

$$(g \circ f)(F) = g\left(\bigcup F\right) = \binom{\bigcup F}{k}$$

$$= \left\{ A \subseteq [n] : |A| = k, A \cap v = \emptyset \text{ for all } v \in \binom{[n] \setminus \bigcup F}{k} \right\}$$

$$= \{ A \subseteq [n] : |A| = k, A \cap v = \emptyset \text{ for all } v \in \nu(F) \}$$

$$= \nu^2(F).$$

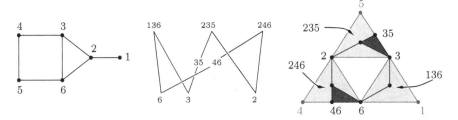

Fig. 2.6 A graph, its poset of closed sets, and the Lovász complex

The Lovász Complex

We will now concentrate on the image of the neighbor set function ν in general and use it to define a new simplicial complex that we will call the Lovász complex, $\mathcal{L}(G)$, of a graph G. It is very closely related to the neighborhood complex, $\mathcal{N}(G)$. In fact, $\mathcal{L}(G)$ will be a strong deformation retract of $\mathcal{N}(G)$ that has more structural properties. The richer structure will allow more topological tools to be used, namely the Borsuk–Ulam theorem.

A set $A \subseteq V$ in the image of ν has the property that $\nu^2(A) = A$. To see this, suppose that $A = \nu(B)$. Then $\nu^2(A) = \nu^3(B) = \nu(B) = A$, by property 3 in Proposition 2.5. We will call sets A with this property *closed*, since ν has the properties of a closure operator. Denote by $C(G)$ the set of all nonempty proper subsets of V that are closed, i.e.,

$$C(G) = \left\{ A \subset V : A \neq \emptyset,\ A \neq V,\ \nu^2(A) = A \right\}.$$

For example, in the case of a Kneser graph $KG_{n,k}$, a set $F \subseteq \binom{[n]}{k}$ of vertices is closed if and only if $\nu^2(F) = \binom{\bigcup F}{k} = F$.

Let $G = (V, E)$ be a finite simple graph. Let the *Lovász complex* $\mathcal{L}(G)$ be the order complex of the partially ordered set $(C(G), \subseteq)$, i.e., the order complex of all nonempty proper closed subsets of V ordered by inclusion. In particular, note that each element of $C(G)$ is a face of the neighborhood complex $\mathcal{N}(G)$, and hence $\mathcal{L}(G)$ will be a subcomplex of the first barycentric subdivision of the neighborhood complex. For the example graph in Fig. 2.3, the graph itself, the poset of closed sets, and the Lovász complex as a subcomplex of sd $\mathcal{N}(G)$ are shown in Fig. 2.6. Note how the neighbor set function ν acts on this complex.

In particularly nice situations, every face of $\mathcal{N}(G)$ is closed. One such case is that of the complete graph K_n on the vertex set $[n]$. In this case $\nu(A) = [n] \setminus A$. As we have seen already, $\mathcal{N}(K_n)$ is the boundary complex of an $(n-1)$-dimensional simplex. Hence, $\mathcal{L}(G)$ is the order complex of its face poset, and therefore the barycentric subdivision of the simplex boundary. Figure 2.7 shows the neighborhood

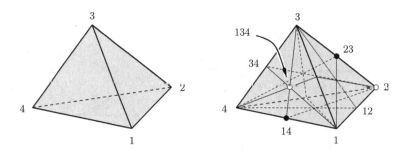

Fig. 2.7 The neighborhood and Lovász complexes of K_4

and Lovász complexes for the complete graph K_4. It also illustrates that the neighbor set function ν is topologically the antipodal map on $\mathcal{L}(K_n)$, i.e., for example $\nu(\{2\}) = \{134\}$ and $\nu(\{1, 2\}) = \{3, 4\}$.

Proposition 2.6. *There is a homeomorphism* $\varphi : |\mathcal{L}(K_n)| \to \mathbb{S}^{n-2}$ *that is* $\mathbb{Z}_2$-*equivariant, i.e.,* $\varphi(\nu(x)) = -\varphi(x)$.

Proof. We may geometrically realize $\mathcal{L}(K_n)$ as the boundary of the standard $(n-1)$-dimensional simplex whose points are given by convex combinations $\sum_{i=1}^{n} t_i e_i$, with $t_i \geq 0$, $\sum_{i=1}^{n} t_i = 1$, and $t_i = 0$ for at least one i. A vertex $A \subset [n]$ of $\mathcal{L}(K_n)$ then corresponds to the point $e_A = \frac{1}{|A|} \sum_{i \in A} e_i$. Note that by definition, the induced action on these vertices is given by $\nu(e_A) = e_{\nu(A)}$. Moreover, observe that we may identify the sphere $\mathbb{S}^{n-2}$, together with its antipodal action, with the subspace $S = \mathbb{S}^{n-1} \cap \{x : \sum_{i=1}^{n} x_i = 0\} \subseteq \mathbb{R}^n$. We will construct an equivariant homeomorphism $\varphi : |\mathcal{L}(K_n)| \to S$ in two steps. We will first define φ on the points corresponding to vertices of $\mathcal{L}(K_n)$, and then extend the map predetermined by the $\mathbb{Z}_2$-equivariance.

Let's denote the center of the $(n-1)$-dimensional standard simplex by $c = \left(\frac{1}{n}, \ldots, \frac{1}{n}\right)$. Then define φ for any $A \subset [n]$ by

$$e_A \longmapsto \frac{e_A - c}{\|e_A - c\|}.$$

We claim that φ is $\mathbb{Z}_2$-equivariant on the set of points e_A, $A \subset [n]$. In order to show this, it suffices to show that c lies on the line segment between e_A and $\nu(e_A)$. But clearly

$$c = \frac{|A|}{n} e_A + \left(1 - \frac{|A|}{n}\right) \nu(e_A),$$

since $(1 - \frac{|A|}{n}) = \frac{|\nu(A)|}{n}$. We may now extend the map to all of $|\mathcal{L}(K_n)|$. An arbitrary point of $|\mathcal{L}(K_n)|$ is given by $\sum_{i=1}^{k} t_i e_{A_i}$ for some chain $A_1 \subset \cdots \subset A_k \subset [n]$ and $t_i \geq 0$, $\sum_{i=1}^{k} t_i = 1$. Extending φ by

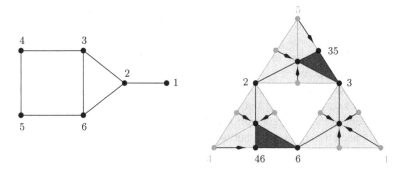

Fig. 2.8 The retraction given by ν^2 in the proof of Proposition 2.7

$$\varphi\left(\sum_{i=1}^{k} t_i e_{A_i}\right) = \frac{\sum_{i=1}^{k} t_i \varphi(e_{A_i})}{\|\sum_{i=1}^{k} t_i \varphi(e_{A_i})\|}$$

yields a continuous map, which is equivariant by definition. It is left to the reader to show the bijectivity of the resulting map. $\qquad\square$

For an alternative proof of the previous proposition we refer to Exercise 8 on page 142.

Proposition 2.7. *The Lovász complex* $\mathcal{L}(G)$ *is a strong deformation retract of the neighborhood complex* $\mathcal{N}(G)$. *In particular, the two complexes are homotopy equivalent.*

Proof. Consider the map $\nu^2 : \mathcal{F}(\mathcal{N}(G)) \to \mathcal{F}(\mathcal{N}(G))$. As remarked before, since $\nu^3 = \nu$, the image of this map is $C(G)$. Let $i : C(G) \to \mathcal{F}(\mathcal{N}(G))$ be the inclusion map. Then $\nu^2 \circ i = \mathrm{id}_{C(G)}$, and $(i \circ \nu^2)(A) \supseteq A = \mathrm{id}_{\mathcal{F}(\mathcal{N}(G))}(A)$ for all $A \in \mathcal{F}(\mathcal{N}(G))$. Hence, by the order homotopy lemma, Lemma C.3, the order complex $\Delta(C(G))$ is a strong deformation retract of $\Delta(\mathcal{F}(\mathcal{N}(G)))$, which in turn is the first barycentric subdivision of $\mathcal{N}(G)$. The retraction map ν^2 for our example graph is illustrated in Fig. 2.8. $\qquad\square$

A $\mathbb{Z}_2$-Action on $\mathcal{L}(G)$

We will now turn our attention to the richer structural properties of the Lovász complex $\mathcal{L}(G)$. This was first investigated by James Walker [Wal83].

By statement 3 of Proposition 2.5, the map ν induces a bijective simplicial map from $\mathcal{L}(G)$ to itself. First of all, it is a self-inverse bijection of the vertices, and furthermore it is order-reversing. Hence, ν maps inclusion chains to inclusion chains, i.e., simplices to simplices.

Since $\nu^2 = $ id on $\mathcal{L}(G)$ we may identify the pair $\{$id$, \nu\}$ with the 2-element group $\mathbb{Z}_2$. Note that ν leaves no $A \subseteq V$ fixed, and therefore ν provides a free $\mathbb{Z}_2$-action on the Lovász complex.

We know already that in the case of the Kneser graphs we have homotopy equivalences $|\mathcal{L}(KG_{n,k})| \simeq |\mathcal{N}(KG_{n,k})| \simeq |\Delta(B_{n,k})|$. Now observe that $B_{n,k}$ is also equipped with a fixed-point-free, order-reversing involution μ given by taking complements, i.e., $\mu(A) = [n] \setminus A$ for $A \in B_{n,k}$.

The content of the next proposition is that these two $\mathbb{Z}_2$-actions are compatible. We will not need it in the sequel, but we discuss it in order to get more accustomed to the Lovász complex and the neighbor set function ν.

Proposition 2.8. *There is a $\mathbb{Z}_2$-equivariant homeomorphism from $\mathcal{L}(KG_{n,k})$ to the order complex $\Delta(B_{n,k})$.*

Proof. We are essentially proving a stronger form of Proposition 2.4 along the same lines. Let $f' : C(G) \rightarrow B_{n,k}$ be defined by $F \mapsto \bigcup F$ and let $g' : B_{n,k} \rightarrow C(G)$ be given by $A \mapsto \binom{A}{k}$. Then $f' \circ g' = $ id$_{B_{n,k}}$ is clear, and $(g' \circ f')(F) = \nu^2(F) = F = $ id$_{C(G)}(F)$ for all $F \in C(G)$. Both maps, f' and g', are order-preserving, and hence the order complexes of both posets are homeomorphic.

Now compute

$$f'(\nu(F)) = f'(\nu(\binom{\bigcup F}{k})) = f'(\binom{[n] \setminus \bigcup F}{k}) = [n] \setminus \bigcup F = \mu(f'(F)),$$

which yields the $\mathbb{Z}_2$-equivariance.   □

Note that the map f in the proof of Proposition 2.4 actually factors through $C(G)$, as $f = f' \circ \nu^2$.

Graph Homomorphisms and Induced Maps

We now turn our attention to the proof of Lovász's theorem, Theorem 2.3. An important property of the constructions of the neighborhood and Lovász complexes is the property that every graph homomorphism yields a simplicial map of the associated complexes.

Let $G = (V, E)$ and $H = (V', E')$ be two finite simple graphs with neighbor set functions ν and ν'. Let $f : G \rightarrow H$ be a graph homomorphism, i.e., a map $f : V \rightarrow V'$ with the property that for every edge $uv \in E$, the image $f(u)f(v)$ is an edge of H.

We will abuse notation and write $f(A)$ for the image of a subset $A \subseteq V$ under f. As Fig. 2.9 demonstrates, the inclusion $f(\nu(A)) \subseteq \nu'(f(A))$ holds for all $A \subseteq V$.

This implies, in particular, that if $\nu(A) \neq \emptyset$, then $\nu'(f(A)) \supseteq f(\nu(A)) \neq \emptyset$. Hence, f induces a simplicial map $\mathcal{N}(f) : \mathcal{N}(G) \rightarrow \mathcal{N}(H)$ by $A \mapsto f(A)$.

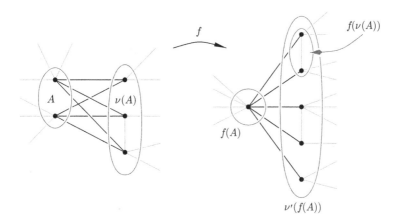

Fig. 2.9 The interplay between the neighbor set functions and graph homomorphisms

As Fig. 2.9 shows, in general the inclusion $f(v(A)) \subseteq v'(f(A))$ can be proper. Hence, in order to obtain an induced map on the Lovász complex, we have to take the closure of the image under f. The induced map $\mathcal{L}(f) : \mathcal{L}(G) \to \mathcal{L}(H)$ is given on the vertices of $\mathcal{L}(G)$ by $A \mapsto (v')^2(f(A))$. It yields a simplicial map by the basic observation that $(v')^2(f(A)) \subseteq (v')^2(f(B))$ for any $A \subseteq B \subseteq V$. The map $\mathcal{L}(f)$ respects the neighbor set functions, i.e., $\mathcal{L}(f)$ is equivariant with respect to the $\mathbb{Z}_2$-actions that v and v' induce. In order to see this we need the following lemma.

Lemma 2.9. *For any graph homomorphism $f : G \to H$, the following relation holds for any closed set $A \in C(G)$:*

$$(v')^2(f(v(A)) = v'(f(A)).$$

Proof. We know already that $f(v(A)) \subseteq v'(f(A))$ for all A. We therefore have

$$(v')^2(f(v(A))) \subseteq (v')^3(f(A)) = v'(f(A))$$

and

$$v'(f(A)) = v'(f(v^2(A))) = v'(f(v(v(A)))) \subseteq (v')^2(f(v(A))). \qquad \square$$

Now consider an inclusion chain $\{A_0 \subset \cdots \subset A_k\}$ of closed sets of G, i.e., a k-simplex of the Lovász complex $\mathcal{L}(G)$. Then

$$
\begin{aligned}
\mathcal{L}(f)(v(\{A_0 \subset \cdots \subset A_k\})) &= \mathcal{L}(f)(\{v(A_k) \subset \cdots \subset v(A_0)\}) \\
&= \{(v')^2(f(v(A_k))) \subset \cdots \subset (v')^2(f(v(A_0)))\} \\
&= \{v'(f(A_k)) \subset \cdots \subset v'(f(A_0))\} \\
&= v'(\{f(A_0) \subset \cdots \subset f(A_k)\}) \\
&= v'(\mathcal{L}(f)(\{A_0 \subset \cdots \subset A_k\})).
\end{aligned}
$$

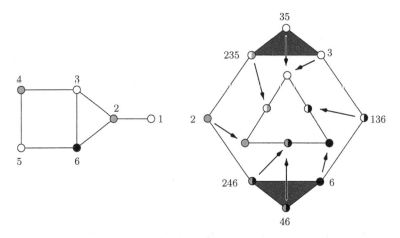

Fig. 2.10 A three-coloring of G and the induced map of the Lovász complexes

We summarize the previous insights.

Proposition 2.10. *Any graph homomorphism* f : G → H *induces a* $\mathbb{Z}_2$-*equivariant simplicial map* $\mathcal{L}(f)$: $\mathcal{L}(G)$ → $\mathcal{L}(H)$.   □

Before we apply this to prove Theorem 2.3, let's consider an enlightening example. The graph homomorphisms we are mostly interested in are induced by a coloring of the graph. In fact, if c : $V(G)$ → $[m]$ is a proper m-coloring of the graph, then c induces a graph homomorphism G → K_m. And conversely, every graph homomorphism G → K_m yields an m-coloring of the graph. In other words, the chromatic number of a graph G equals

$$\chi(G) = \min\{m \geq 0 : \text{there exists a graph homomorphism } G \to K_m\}.$$

As an example, we consider a three-coloring of our example graph. We color the vertices with colors white, gray, and black. Figure 2.10 shows the coloring along with the induced map $\mathcal{L}(G) \to \mathcal{L}(K_3)$.

Proof (of Theorem 2.3). Assume that G possesses a proper m-coloring, i.e., there exists a graph homomorphism f : G → K_m. Since $\mathcal{N}(G)$ is k-connected by assumption, so is $\mathcal{L}(G)$ by Proposition 2.7. Hence, there exists a $\mathbb{Z}_2$-equivariant map $\psi : \mathbb{S}^{k+1} \to |\mathcal{L}(G)|$, where $\mathbb{Z}_2$ acts on $\mathbb{S}^{k+1}$ via the antipodal map. Such a map can easily be constructed inductively using a $\mathbb{Z}_2$-invariant triangulation of the sphere such as, for example, that given by the boundary complex of the cross polytope. The details are given in the proof of Proposition D.13 on page 216. Together with Proposition 2.6, we obtain the following composition of $\mathbb{Z}_2$-equivariant maps:

$$\mathbb{S}^{k+1} \xrightarrow{\psi} |\mathcal{L}(G)| \xrightarrow{|\mathcal{L}(f)|} |\mathcal{L}(K_m)| \xrightarrow{\varphi} \mathbb{S}^{m-2}.$$

By the Borsuk–Ulam theorem we have $m - 2 \geq k + 1$, and hence $m \geq k + 3$.   □

Finally, we obtain a proof of Kneser's conjecture along the lines of Lovász's original proof.

Corollary 2.11. *For the family of Kneser graphs $KG_{n,k}$ we obtain $\chi(KG_{n,k}) \geq n - 2k + 2$.*

Proof. By Theorem 2.3 and Proposition 2.4 we obtain

$$\chi(KG_{n,k}) \geq \operatorname{conn}(|\mathcal{N}(KG_{n,k})|) + 3 = n - 2k - 1 + 3 = n - 2k + 2. \qquad \square$$

2.3 A Conjecture by Lovász

This section is devoted to a more recent development. It is about a general approach to endowing the category of graphs with topological structure, and in fact can be seen as a generalization of the concepts we discussed in the previous sections of this chapter. The concept was introduced by László Lovász, and the story line develops along a conjecture by him claiming a somewhat analogous statement to Theorem 2.3. The conjecture was proved by Eric Babson and Dmitry Kozlov in 2005 [BK07, Koz07]. A shorter and very elegant proof was later found by Carsten Schultz [Schu06]. We will present his argument and follow in many respects his original article.

By the definition of the neighbor set function ν, which assigns the common neighbors to a set of vertices in a graph, pairs $A, \nu(A)$ are the shores of complete bipartite subgraphs. What does it mean for two sets $A, B \subseteq V$ to be the two shores of a complete bipartite subgraph of G? A fancy way to say it is that every choice of vertices $u \in A$ and $v \in B$ induces a graph homomorphism $\varphi : K_2 \to G$ defined by $\varphi(0) = u$ and $\varphi(1) = v$. Compare Fig. 2.11. In terms of the neighbor set function ν, this amounts to requiring that $A \subseteq \nu(B)$. Note that this implies $B \subseteq \nu^2(B) \subseteq \nu(A)$.

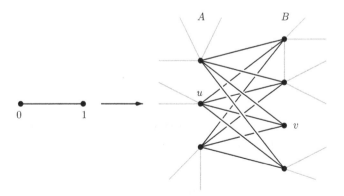

Fig. 2.11 Graph homomorphisms $K_2 \to G$ and shores of bipartite subgraphs

Hom *Complexes*

The interpretation above leads to the following generalization of graph homomorphisms. Let $T = (V', E')$ and $G = (V, E)$ be graphs. A *multihomomorphism* from
T to G is a map $\varphi : V' \to \mathcal{P}(V) \setminus \{\emptyset\}$ associating a nonempty subset of V to every
vertex of T such that every function $f : V' \to V$, with $f(v) \in \varphi(v)$ for all $v \in V$,
is a graph homomorphism from T to G.

Note that indeed, a multihomomorphism has the property that for any pair
$u, v \in V'$ of adjacent vertices, the sets $\varphi(u)$ and $\varphi(v)$ are the vertex sets of complete
bipartite subgraphs of G. In particular, any multihomomorphism from K_2 to a graph
G is given by nonempty sets $A, B \subset V$ that are vertex sets of a complete bipartite
subgraph of G.

Each multihomomorphism φ from T to G can be identified with a product of
geometric simplices contained in

$$\prod_{u \in V(T)} \Delta_{|V(G)|-1}$$

as follows. We clearly may identify the subsets of $V(G)$ with the faces of $\Delta_{|V(G)|-1}$.
Then, for each $u \in V(T)$, the subset $\varphi(u) \subseteq V(G)$ defines a face $F_u^\varphi \subseteq \Delta_{|V(G)|-1}$.
With the multihomomorphism φ, we now associate the product

$$\prod_{u \in V(T)} F_u^\varphi \subseteq \prod_{u \in V(T)} \Delta_{|V(G)|-1}.$$

We denote the set of multihomomorphisms from T to G by $\mathrm{Hom}(T, G)$ and
denote its *geometric realization* by

$$|\mathrm{Hom}(T, G)| = \bigcup_{\substack{\varphi : V(T) \to \mathcal{P}(V(G)) \setminus \{\emptyset\} \\ \text{multihom.}}} \left(\prod_{u \in V(T)} F_u^\varphi \right) \subseteq \prod_{u \in V(T)} \Delta_{|V(G)|-1}.$$

First Examples

As a first example, consider $\mathrm{Hom}(K_2, C_3)$, where $C_3 = \triangle$ is a cycle of length 3. The
multihomomorphisms in this case are given by all ordered pairs (A, B) describing
two shores of a complete bipartite subgraph. If we denote the vertices of C_3 by
$0, 1, 2$, these pairs are

$$(0, 1), (0, 2), (1, 0), (1, 2), (2, 0), (2, 1),$$

$$(01, 2), (02, 1), (12, 0), (0, 12), (1, 02), (2, 01),$$

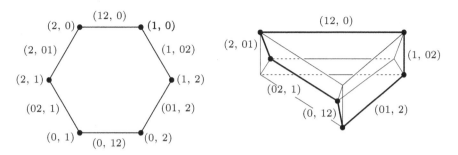

Fig. 2.12 $|\text{Hom}(K_2, C_3)|$ by itself and as a subcomplex of $\Delta_2 \times \Delta_2$

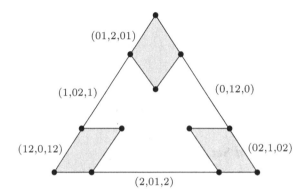

Fig. 2.13 The space $|\text{Hom}(P_2, C_3)|$

where we have abused notation by writing $(0, 1)$ instead of $(\{0\}, \{1\})$ and $(01, 2)$ instead of $(\{0, 1\}, \{2\})$, etc. The geometric realization of $\text{Hom}(K_2, C_3)$ consists of six edges forming a circle. Figure 2.12 shows $|\text{Hom}(K_2, C_3)|$ by itself and as a subcomplex of the product

$$\prod_{u \in V(K_2)} \Delta_{|V(C_3)|-1} = \Delta_2 \times \Delta_2.$$

As with the Lovász complex $\mathcal{L}(C_{2r+1})$ of an odd cycle, $|\text{Hom}(K_2, C_{2r+1})|$ is always homeomorphic to a circle, i.e., a 1-dimensional sphere, and $|\text{Hom}(K_2, C_{2r})|$ is homeomorphic to two disjoint circles.

The next example, $\text{Hom}(P_2, C_3)$, for $P_2 = $ •—•—• a path of length 2, involves higher-dimensional cells. Each multihomomorphism is now given by a triple (A, B, C) such that each of the pairs (A, B) and (B, C) are the shores of complete bipartite subgraphs of C_3. Typical examples of dimension one and two are $(0, 12, 0)$ and $(12, 0, 12)$. Figure 2.13 shows the space $|\text{Hom}(P_2, C_3)|$ with the 1- and 2-dimensional cells labeled.

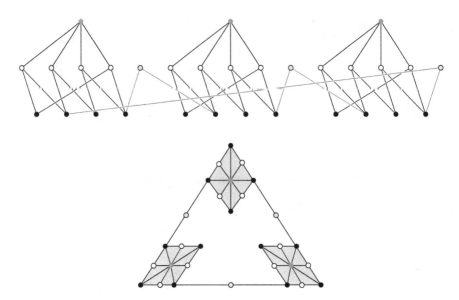

Fig. 2.14 The poset $(\mathrm{Hom}(P_2, C_3), \subseteq)$ and the order complex $\Delta(\mathrm{Hom}(P_2, C_3))$

The Partially Ordered Set Structure

The set $\mathrm{Hom}(T, G)$ of multihomomorphisms from T to G is partially ordered by inclusion, i.e., $\varphi \subseteq \psi$ holds for $\varphi, \psi \in \mathrm{Hom}(T, G)$ if and only if $\varphi(u) \subseteq \psi(u)$ for all $u \in V(T)$. The order corresponds to the partial inclusion order of cells

$$\prod_{u \in V(T)} F_u^{\varphi} \subseteq |\mathrm{Hom}(T, G)| , \; \varphi \in \mathrm{Hom}(T, G).$$

We will always refer to this partial order when considering $\mathrm{Hom}(T, G)$ as a partially ordered set.

In particular, we can assign the order complex to this partial order and obtain a simplicial complex $\Delta(\mathrm{Hom}(T, G))$ whose geometric realization $|\Delta(\mathrm{Hom}(T, G))|$ is homeomorphic to $|\mathrm{Hom}(T, G)|$. This is the content of Exercise 11 on page 68. Figure 2.14 shows the Hasse diagram of the poset $(\mathrm{Hom}(P_2, C_3), \subseteq)$ and the order complex $\Delta(\mathrm{Hom}(P_2, C_3))$.

Just as we can compose graph homomorphisms, we can do so with graph multihomomorphisms. We suggestively use the following notation:

$$\mathrm{Hom}(T, G) \times \mathrm{Hom}(G, H) \longrightarrow \mathrm{Hom}(T, H),$$

$$(\varphi, \psi) \longmapsto \varphi * \psi,$$

where the latter is defined by

$$(\varphi * \psi)(u) = \{w \in V(H) : \text{there exists } v \in \varphi(u) \text{ such that } w \in \psi(v)\}.$$

It is an easy exercise to see that $\varphi * \psi$ is a graph multihomomorphism from T to H that respects the inclusion order, i.e., $\varphi \subseteq \varphi'$ and $\psi \subseteq \psi'$ implies $\varphi * \psi \subseteq \varphi' * \psi'$. This implies that $*$ induces a simplicial map on the order complex. Compare Exercise 12.

Proposition 2.12. *For any three graphs T, G, H there is a continuous map*

$$* : |\text{Hom}(T, G)| \times |\text{Hom}(G, H)| \longrightarrow |\text{Hom}(T, H)|,$$

which, restricted to graph homomorphisms, is identical to ordinary composition. Moreover, $*$ satisfies the associativity law.

Proof. The result follows from Lemma C.2, i.e., the fact that there is a homeomorphism $|\Delta(P \times Q)| \cong |\Delta(P)| \times |\Delta(Q)|$. The details are left to Exercise 13. $\square$

$\mathbb{Z}_2$-Structure

As in the previous section, we need more structure (such as a free $\mathbb{Z}_2$-action) on the spaces we are considering. Note that $T = K_2$, as well as any cycle $T = C_n$, admits a self-inverse automorphism flipping an edge, i.e., a graph isomorphism $\gamma : T \to T$ with $\gamma^2 = \text{id}_T$ and such that there exists an edge $uv \in E(T)$ with $\gamma(u) = v$ (and hence $\gamma(v) = u$). Let K_2 and the odd cycle C_{2r+1} have vertex sets $\{0, 1\}$ and $\{0, 1, \ldots, 2r\}$, respectively. Denote by $\alpha : K_2 \to K_2$ the automorphism given by $\alpha(i) = 1-i$, and by $\beta : C_{2r+1} \to C_{2r+1}$ the automorphism given by $\beta(i) = 2r - i$. See Fig. 2.15.

Proposition 2.13. *Any self-inverse graph automorphism $\gamma : T \to T$ flipping an edge induces a free $\mathbb{Z}_2$-action on $|\text{Hom}(T, G)|$ for any graph G via*

$$|\text{Hom}(T, G)| \longrightarrow |\text{Hom}(T, G)|,$$

$$x \longmapsto \gamma * x.$$

Proof. It is clear that the map is self-inverse, since $\gamma * (\gamma * x) = (\gamma * \gamma) * x = x$. We need to check that it is fixed-point-free. Assume that the edge flipped by γ has vertices u and v. If $x \in \prod_{w \in V(T)} F_w^\varphi \subseteq |\text{Hom}(T, G)|$, then clearly $\gamma * x \in \prod_{w \in V(T)} F_w^{\gamma * \varphi}$. Now

$$F_u^\varphi \cap F_u^{\gamma * \varphi} = F_u^\varphi \cap F_v^\varphi = \emptyset,$$

and hence $\gamma * x \neq x$. $\square$

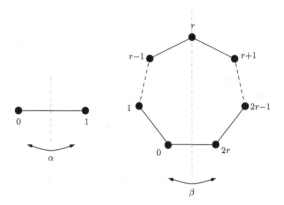

Fig. 2.15 The edge-flipping automorphisms α of K_2 and β of C_{2r+1}

Corollary 2.14. *For any graph* G, *the map* α *induces a free* $\mathbb{Z}_2$-*action on* $|\operatorname{Hom}(K_2, G)|$ *and* β *does so on* $|\operatorname{Hom}(C_{2r+1}, G)|$.   □

Lovász and Hom *Complexes*

As we already discussed, the partially ordered set $\operatorname{Hom}(K_2, G)$ may be identified with the set of all pairs (A, B), with $A, B \subseteq V(G)$ nonempty sets that are the two shores of a complete bipartite graph ordered by componentwise inclusion. The correspondence is given by $\varphi(0) = A$ and $\varphi(1) = B$. Now $(\alpha * \varphi)(0) = \varphi(\alpha(0)) = \varphi(1) = B$, and similarly, $(\alpha * \varphi)(1) = A$. Hence, on the set of pairs the $\mathbb{Z}_2$-action is given by $(A, B) \mapsto (B, A)$. In particular, if we consider a pair $(A, \nu(A))$ for a closed set $A \subseteq V(G)$, we obtain $(A, \nu(A)) \mapsto (\nu(A), A)$. Hence, in the first coordinate we are left with the $\mathbb{Z}_2$-action of ν on the set $C(G)$ of closed subsets.

After these considerations we might expect a close relationship between the Lovász complex $\mathcal{L}(G) = \Delta C(G)$ and $\operatorname{Hom}(K_2, G)$.

The poset $\operatorname{Hom}(K_2, G)$ has been introduced before. Its order complex is a version of the *box complex* and was thoroughly investigated by Matoušek and Ziegler [MZ04]. One of their results is the following.

Proposition 2.15. *There exist simplicial* $\mathbb{Z}_2$-*maps*

$$f : \operatorname{sd} \mathcal{L}(G) \to \Delta \operatorname{Hom}(K_2, G)$$

and $g : \operatorname{sd}(\Delta \operatorname{Hom}(K_2, G)) \to \operatorname{sd} \mathcal{L}(G)$.

Proof. A vertex of $\operatorname{sd} \mathcal{L}(G)$ is given by an inclusion chain $\{A_0 \subset A_1 \subset \cdots \subset A_k\}$ of nonempty closed sets of vertices of the graph G. The $\mathbb{Z}_2$-action on these chains is induced by the $\mathbb{Z}_2$-action on $\mathcal{L}(G)$, and is given by

$$\nu(\{A_0 \subset A_1 \subset \cdots \subset A_k\}) = \{\nu(A_k) \subset \cdots \subset \nu(A_1) \subset \nu(A_0)\}.$$

Define $f : \operatorname{sd} \mathcal{L}(G) \to \Delta \operatorname{Hom}(K_2, G)$ on the vertices by

$$f(\{A_0 \subset A_1 \subset \cdots \subset A_k\}) = (A_0, \nu(A_k)).$$

Then f is well defined, since $A_0 \subseteq \nu(\nu(A_k)) = A_k$ and $\nu(A_k) \subseteq \nu(A_0)$. It is $\mathbb{Z}_2$-equivariant, since

$$f(\nu(\{A_0 \subset A_1 \subset \cdots \subset A_k\})) = \big(\nu(A_k), \nu(\nu(A_0))\big) = (\nu(A_k), A_0).$$

In order to verify that f is simplicial, we consider two chains

$$\{A_0 \subset A_1 \subset \cdots \subset A_k\} \subseteq \{A_0' \subset A_1' \subset \cdots \subset A_l'\}.$$

In this case, $A_0' \subseteq A_0 \subset A_k \subseteq A_l'$ must hold, and therefore $A_0' \subseteq A_0$ and $\nu(A_l') \subseteq \nu(A_k)$.

Let us now construct the map $g : \operatorname{sd}(\Delta \operatorname{Hom}(K_2, G)) \to \operatorname{sd} \mathcal{L}(G)$. A vertex c of $\operatorname{sd}(\Delta \operatorname{Hom}(K_2, G))$ is given by an inclusion chain of pairs $c = \{(A_0, B_0) \subset (A_1, B_1) \subset \cdots \subset (A_k, B_k)\}$ with the property $A_i \subseteq \nu(B_i)$ and $B_i \subseteq \nu(A_i)$, for $i = 0, \ldots, k$. Consider the chain of inclusions

$$\nu^2(A_0) \subseteq \nu^2(A_1) \subseteq \cdots \subseteq \nu^2(A_k) \subseteq \nu^3(B_k) = \nu(B_k) \subseteq \nu(B_{k-1}) \subseteq \cdots \subseteq \nu(B_0)$$

of nonempty closed sets. Define $g(c)$ to be the inclusion chain that one obtains by eliminating repeated sets in this chain. This map is easily seen to be simplicial and $\mathbb{Z}_2$-equivariant.   □

In fact, it is not hard to show that $|\mathcal{L}(G)|$ and $|\operatorname{Hom}(K_2, G)|$ are $\mathbb{Z}_2$-homotopy equivalent. We will discuss this at the end of the section.

By applying Proposition 2.6 we obtain the following corollary.

Corollary 2.16. *There are $\mathbb{Z}_2$-equivariant maps $\mathbb{S}^{n-2} \to |\operatorname{Hom}(K_2, K_n)| \to \mathbb{S}^{n-2}$.*   □

Along the lines of the proof of Lovász's theorem, Theorem 2.3, we obtain a new version of Lovász's theorem.

Corollary 2.17. *For any graph G, there is the following bound on the chromatic number:*

$$\chi(G) \geq \operatorname{conn} |\operatorname{Hom}(K_2, G)| + 3.$$

Proof. Assume that $|\operatorname{Hom}(K_2, G)|$ is k-connected and that there exists an m-coloring of G, i.e., a graph homomorphism $f : G \to K_m$. Then there exists the following sequence of $\mathbb{Z}_2$-maps:

$$\mathbb{S}^{k+1} \longrightarrow |\operatorname{Hom}(K_2, G)| \longrightarrow |\operatorname{Hom}(K_2, K_m)| \longrightarrow \mathbb{S}^{m-2}.$$

By the Borsuk–Ulam theorem we obtain $m - 2 \geq k + 1$.   □

We now turn our attention to the $\mathbb{Z}_2$-homotopy equivalence of $|\mathcal{L}(G)|$ and $|\operatorname{Hom}(K_2, G)|$. The following proof is due to Schultz [Schu10].

In the first step we will replace $\mathcal{L}(G)$ by a $\mathbb{Z}_2$-homeomorphic copy. To this end, recall the construction of the interval order of a partially ordered set as introduced on page 203. We will apply it to the partially ordered set $(C(G), \subseteq)$. In this case,

$$\operatorname{Int}(C(G)) = \{(A, B) : A, B \in C(G), A \subseteq B\}$$

ordered by $(A, B) \preceq (A', B')$ if and only if $A \subseteq A'$ and $B' \subseteq B$.

By Proposition C.5, the geometric realizations of $\mathcal{L}(G) = \Delta(C(G))$ and $\Delta \operatorname{Int}(C(G))$ are homeomorphic. It is an easy exercise to see that under this homeomorphism the $\mathbb{Z}_2$-action on $\operatorname{Int}(C(G))$ defined by $(A, B) \mapsto (\nu(B), \nu(A))$ corresponds to the $\mathbb{Z}_2$-action on $\mathcal{L}(G)$.

Hence, it suffices to prove the following proposition.

Proposition 2.18. *The partial orders* $\operatorname{Int}(C(G))$ *and* $\operatorname{Hom}(K_2, G)$ *are* $\mathbb{Z}_2$-*homotopy equivalent.*

Proof. Define the maps

$$f : \operatorname{Hom}(K_2, G) \longrightarrow \operatorname{Int}(C(G)),$$

$$(A, B) \longmapsto (\nu^2(A), \nu(B)),$$

and

$$g : \operatorname{Int}(C(G)) \longrightarrow \operatorname{Hom}(K_2, G),$$

$$(A, B) \longmapsto (A, \nu(B)).$$

These maps are easily seen to be order-preserving and $\mathbb{Z}_2$-equivariant. Now

$$(f \circ g)(A, B) = f(A, \nu(B)) = (\nu^2(A), \nu^2(B)) = (A, B)$$

and

$$(g \circ f)(A, B) = g(\nu^2(A), \nu(B)) = (\nu^2(A), \nu^2(B)) \supseteq (A, B).$$

In other words, $f \circ g = \operatorname{id}_{\operatorname{Int}(C(G))}$ and $g \circ f \geq \operatorname{id}_{\operatorname{Hom}(K_2, G)}$ with respect to the order on $\operatorname{Hom}(K_2, G)$. By the order homotopy lemma, Lemma C.3 and the following Remark C.4, the composition $g \circ f$ is $\mathbb{Z}_2$-homotopic to the identity.   □

Note that in fact, the previous proof yields that the map g is injective and that the geometric realization of its image is a strong deformation retract of $|\operatorname{Hom}(K_2, G)|$.

Lovász's Conjecture

Contemplating Corollary 2.17, it seems natural to replace K_2 by some other graph in order to obtain new, and maybe stronger, lower bounds for the chromatic number. The existence of a graph homomorphism from K_2 to G witnesses a lower bound of two for the chromatic number, and the existence of a homomorphism of an odd cycle C_{2r+1} to G a lower bound of three. This raises the question whether $\operatorname{Hom}(C_{2r+1}, G)$ can also be used to obtain a general lower bound for the chromatic number. Lovász conjectured that

$$\chi(G) \geq \operatorname{conn}|\operatorname{Hom}(C_{2r+1}, G)| + 4.$$

We will show something slightly stronger. In order to do so we will need a measure for the topological complexity of a space with $\mathbb{Z}_2$-action resembling that of connectivity, but taking the $\mathbb{Z}_2$-action into account.

Definition 2.19. Let X be a topological space with a $\mathbb{Z}_2$-action. The $\mathbb{Z}_2$-*index of* X is defined to be

$$\operatorname{ind}(X) = \min\{k \geq 0 : \text{there exists a continuous } \mathbb{Z}_2\text{-map } X \to \mathbb{S}^k\},$$

i.e., the smallest dimension k such that X can be mapped equivariantly to the k-dimensional sphere endowed with the antipodal action.

Our main example of a space with $\mathbb{Z}_2$-action is the sphere $\mathbb{S}^n$ with the antipodal action. It has $\operatorname{ind}(\mathbb{S}^n) = n$ by the Borsuk–Ulam theorem, Theorem 1.6.

Definition 2.20. Let X be a topological space with a $\mathbb{Z}_2$-action. The $\mathbb{Z}_2$-*coindex of* X is defined to be

$$\operatorname{co-ind}(X) = \max\{k \geq 0 : \text{there exists a continuous } \mathbb{Z}_2\text{-map } \mathbb{S}^k \to X\},$$

i.e., the largest dimension k such that the k-dimensional sphere endowed with the antipodal action can be mapped equivariantly to X.

Again for our main example we have, by the Borsuk–Ulam theorem, Theorem 1.6, that $\operatorname{co-ind}(\mathbb{S}^n) = n$.

Lemma 2.21. *Let X be a topological space with a free $\mathbb{Z}_2$-action. Then the following inequalities hold:*

$$\operatorname{ind}(X) \geq \operatorname{co-ind}(X) \geq \operatorname{conn}(X) + 1.$$

Proof. Let $k = \text{co-ind}(X)$ and $l = \text{ind}(X)$. Then there exists the following composition of $\mathbb{Z}_2$-equivariant maps:

$$\mathbb{S}^k \longrightarrow X \longrightarrow \mathbb{S}^l.$$

And hence, by the Borsuk–Ulam theorem, $\text{ind}(X) - l \geq k = \text{co-ind}(X)$,

Now let $k = \text{conn}(X)$. Then, by Proposition D.13, there exists a $\mathbb{Z}_2$-equivariant map $\mathbb{S}^{k+1} \to X$, and hence $\text{co-ind}(X) \geq k + 1 = \text{conn}(X) + 1$.   □

We will now prove a stronger version of Lovász's theorem, Theorem 2.3, from page 42.

Proposition 2.22. *For any finite simple graph G, the following lower bound holds for the chromatic number:*

$$\chi(G) \geq \text{ind}(|\text{Hom}(K_2, G)|) + 2.$$

Proof. Let $m = \chi(G)$ be the chromatic number of G. Hence, there exists a graph homomorphism $\varphi : G \to K_m$. This yields a map

$$\text{Hom}(K_2, G) \longrightarrow \text{Hom}(K_2, K_m),$$

$$\psi \longmapsto \psi * \varphi,$$

which is $\mathbb{Z}_2$-equivariant. Applying Corollary 2.16, we obtain a $\mathbb{Z}_2$-equivariant map $|\text{Hom}(K_2, G)| \to \mathbb{S}^{m-2}$, and hence

$$\text{ind}(|\text{Hom}(K_2, G)|) + 2 \leq m = \chi(G).$$ □

Hom(K_2, C_{2r+1})

We now turn our attention toward a proof of Lovász's conjecture concerning the relation between $\chi(G)$ and $\text{Hom}(C_{2r+1}, G)$. Since we understand the relation between $\chi(G)$ and $\text{Hom}(K_2, G)$, and we have the map

$$|\text{Hom}(K_2, C_{2r+1})| \times |\text{Hom}(C_{2r+1}, G)| \longrightarrow |\text{Hom}(K_2, G)|,$$

we will first investigate $|\text{Hom}(K_2, C_{2r+1})|$. Figure 2.16 depicts the order complex $\Delta(\text{Hom}(K_2, C_{2r+1}))$. The gray bullets (● and ●) correspond to graph multiho-momorphisms $K_2 \to C_{2r+1}$ that actually are graph homomorphisms such as $\varphi(0) = \{i\}$ and $\varphi(1) = \{i + 1\}$. For each edge e of C_{2r+1}, there are exactly

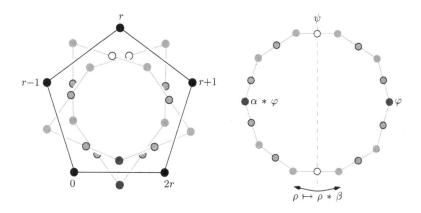

Fig. 2.16 The complex $\Delta(\mathrm{Hom}(K_2, C_{2r+1}))$ for $r = 2$

two homomorphisms mapping K_2 to e. The bullets with a black rim (◉ and ○) correspond to graph multihomomorphisms $\varphi : K_2 \to C_{2r+1}$ such as $\varphi(0) = \{i\}$ and $\varphi(1) = \{i - 1, i + 1\}$.

We want to consider two particular elements $\varphi, \psi \in \mathrm{Hom}(K_2, C_{2r+1})$. They are defined by

$$\varphi(i) = \begin{cases} \{0\}, & \text{if } i = 0, \\ \{2r\}, & \text{if } i = 1, \end{cases} \qquad \psi(i) = \begin{cases} \{r\}, & \text{if } i = 0, \\ \{r - 1, r + 1\}, & \text{if } i = 1, \end{cases}$$

and they are shown in Fig. 2.16 as dark gray bullets. These elements induce continuous maps $f, g : |\mathrm{Hom}(C_{2r+1}, G)| \to |\mathrm{Hom}(K_2, G)|$ via $f(x) = \varphi * x$ and $g(x) = \psi * x$. Let's see how these maps behave with respect to the free $\mathbb{Z}_2$-actions given by β and α, as in Corollary 2.14:

$$f(\beta * x) = \varphi * (\beta * x) = (\varphi * \beta) * x = (\alpha * \varphi) * x = \alpha * (\varphi * x) = \alpha * f(x)$$

$$g(\beta * x) = \psi * (\beta * x) = (\psi * \beta) * x = \psi * x = g(x).$$

Here we used the fact that $\varphi * \beta = \alpha * \varphi$ and $\psi * \beta = \beta$, as is easily checked. In fact, the map $\rho \mapsto \rho * \beta$ corresponds to reflection along the dotted gray line in Fig. 2.16. To summarize, f is equivariant with respect to the $\mathbb{Z}_2$-actions, whereas g is constant on each orbit, i.e., constant on each pair $x, \beta * x$. Moreover, f and g are homotopic, since φ and ψ are connected by a path, i.e., a continuous map $h : [0, 1] \to |\mathrm{Hom}(K_2, C_{2r+1})|$ with $h(0) = \varphi$ and $h(1) = \psi$; see Fig. 2.16. The homotopy from f to g is now given by

$$H : |\mathrm{Hom}(C_{2r+1}, G)| \times [0, 1] \longrightarrow |\mathrm{Hom}(K_2, G)|,$$

$$(x, t) \longmapsto h(t) * x.$$

Let us call $\mathbb{Z}_2$-equivariant maps *odd* and maps that are constant on each orbit *even*.

Lemma 2.23. *Let X, Y be free $\mathbb{Z}_2$-spaces with $Y \neq \emptyset$, and $f, g : X \to Y$ homotopic continuous maps, such that f is odd and g even. Then*

$$\text{ind}(Y) \geq \text{co-ind}(X) + 1.$$

Before we prove the lemma, we will use it to prove Lovász's conjecture. We obtain the following inequality immediately.

Corollary 2.24. *The previously considered maps f and g imply*

$$\text{ind}(|\,\text{Hom}(K_2, G)\,|) \geq \text{co-ind}(|\,\text{Hom}(C_{2r+1}, G)\,|) + 1. \qquad \square$$

The stronger form of Lovász's 1978 theorem, Theorem 2.3, as shown in Proposition 2.22, yields

$$\chi(G) \geq \text{ind}(|\,\text{Hom}(K_2, G)\,|) + 2 \geq \text{co-ind}\,|\,\text{Hom}(C_{2r+1}, G)\,| + 3$$
$$\geq \text{conn}\,|\,\text{Hom}(C_{2r+1}, G)\,| + 4,$$

and we have therefore obtained a proof of Lovász's conjecture that we will state as a theorem in slightly stronger form than conjectured.

Theorem 2.25 (Babson, Kozlov [BK07]). *For any graph G, the inequality $\chi(G) \geq \text{co-ind}\,|\,\text{Hom}(C_{2r+1}, G)\,| + 3$ holds.* $\qquad \square$

The proof of Lemma 2.23 relies on several concepts from the previous chapter. In particular, the strong Ky Fan theorem plays an essential role.

Proof (of Lemma 2.23). Assume to the contrary that $\text{ind}(Y) \leq \text{co-ind}(X)$ and let $k = \text{co-ind}(X)$. Then there exist the compositions of continuous maps

$$\bar{f}, \bar{g} : \mathbb{S}^k \longrightarrow X \xrightarrow{f,g} Y \longrightarrow \mathbb{S}^k,$$

so that by the $\mathbb{Z}_2$-equivariance of the first and last maps, the map $\bar{f}$ is odd, and $\bar{g}$ is even. Since f and g are homotopic, so are $\bar{f}$ and $\bar{g}$. Let $H : \mathbb{S}^k \times [0, 1] \to \mathbb{S}^k$ be the homotopy from $\bar{f}$ to $\bar{g}$.

We need a simplicial version of H with the property that on the boundary the two simplicial maps maintain their parity. This can be done easily. We start with a simplicial version of $\mathbb{S}^k$ by considering the boundary $\partial Q^{k+1} = |\Gamma^k|$ of the $(k + 1)$-dimensional cross polytope. Denote by $\mathcal{F}$ the face poset of the corresponding geometric complex Γ^k, and let $S = \Delta(\mathcal{F} \times \mathcal{I})$, where $\mathcal{I}$ is the poset $(\{0, 1\}, <)$. See Fig. 2.17 for an illustration.

Then $|S| \cong \mathbb{S}^k \times [0, 1]$, and hence we can assume that H is a map from $|S|$ to $|\Gamma^k|$. Now let r be large enough that there exists a simplicial approximation $\mathcal{H} : \text{sd}^r S \to \Gamma^k$, where $\mathcal{H}$ restricted to the two boundary components is odd and even,

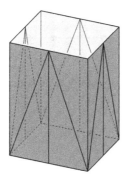

Fig. 2.17 The complex $S = \Delta(\mathcal{F} \times \mathcal{I})$ for $k = 1$

respectively. This can be done easily by making choices consistent with the $\mathbb{Z}_2$-action on the boundary components, since $\bar{f}$ and $\bar{g}$ were odd and even, respectively. Compare the construction of simplicial approximation on page 217 and Exercise 16 on page 68.

By identifying $\pm e_i$ with $\pm i$, the map $\mathcal{H}$ can be considered a labeling function $\text{vert}(\text{sd}^r S) \to \{\pm 1, \ldots, \pm(k+1)\}$. Since there are no edges $\text{conv}(\{+e_i, -e_i\})$ in Γ^k, this labeling does not admit complementary edges. Moreover, note that there are no $(k+1)$-dimensional alternating simplices in $\text{sd}^r S$, since there are too few labels. And therefore, by the theorem of Ky Fan for pseudomanifolds as discussed in Exercise 13 on page 33, the number of $+$-alternating k-dimensional simplices on the boundary is even.

On the boundary component, where the labeling is odd, the number of $+$-alternating k-simplices is odd by the weak version of Ky Fan's theorem, Theorem 1.8. But on the other component, which has an even labeling, obviously the number of $+$-alternating k-simplices is even. Since the sum of an odd and an even number cannot be even, we have reached a contradiction! □

Theorem 2.25 can be phrased in a slightly stronger way using the concept of *cohomological index*, as shown by Schultz [Schu06]. This concept might prove useful because it can actually be computed, in contrast to the difficult determination of indices, coindices, and connectivity.

2.4 Classes with Good Topological Lower Bounds for the Chromatic Number

The family of Kneser graphs is one example of a graph class for which the bound on the chromatic number obtained from the index of the associated spaces $|\text{Hom}(K_2, G)|$ is sharp. It is certainly interesting to see a general scheme to create families of graphs in which this phenomenon occurs. One such scheme is the generalized Mycielski construction.

Fig. 2.18 The graph P_r^0, a path with a loop at 0

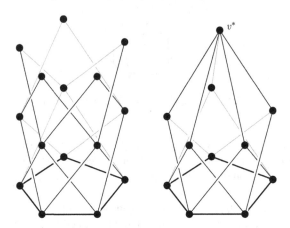

Fig. 2.19 The construction of the Mycielski graph $M_2 C_5$

The Generalized Mycielski Construction

Let G be a finite simple graph and $r \geq 2$ a natural number. We are going to define the *generalized Mycielski graph $M_r G$ of G with parameter r*. To do so, consider the graph P_r^0, a path of length r with vertex set $\{0, 1, \ldots, r\}$ with an additional loop at 0. See Fig. 2.18.

Consider the graph $G \times P_r^0$, which amounts to a copy of G attached to the product of G with a path P_r of length r, i.e.,

$$G \times P_r^0 = G \times \{0\} \cup G \times P_r.$$

The set $V(G) \times \{r\}$ is an independent set in this graph. We will identify all of these vertices in order to obtain $M_r G$, i.e.,

$$M_r G = G \times P_r^0 / (V(G) \times \{r\}).$$

Let us denote the single vertex that is obtained by the identification of $V(G) \times \{r\}$ by v^*. Figure 2.19 shows the two steps of the construction for the 5-cycle $G = C_5$ and $r = 2$, which in this case results in the Grötzsch graph.

This construction increases the chromatic number by at most one, i.e.,

$$\chi(G) + 1 \geq \chi(M_r G).$$

Assume that $c : V(G) \to [k]$ is a proper coloring of G. Then $\bar{c} : V(M_r G) = (V(G) \times \{0, 1, \ldots, r-1\}) \cup \{v^*\} \to [k+1]$ defined by $\bar{c}(v, t) = c(v)$ and $\bar{c}(v^*) = k+1$ is a proper coloring of $M_r G$, as is easily checked. Now the following lemma shows that the topological lower bound on the chromatic number also increases by exactly one.

Lemma 2.26. *Given any finite simple graph G, we have*

$$\text{co-ind}(|\operatorname{Hom}(K_2, M_r G)|) \geq \text{co-ind}(|\operatorname{Hom}(K_2, G)|) + 1.$$

Proof. Let $k = \text{co-ind}(|\operatorname{Hom}(K_2, G)|)$, and let $f : \mathbb{S}^k \to |\operatorname{Hom}(K_2, G)|$ be a $\mathbb{Z}_2$-equivariant map. We have to construct a $\mathbb{Z}_2$-equivariant map $g : \mathbb{S}^{k+1} \to |\operatorname{Hom}(K_2, M_r G)|$. In order to do so, we will work with the following homeomorphic model of $\mathbb{S}^{k+1}$. We know from Appendix B that

$$\mathbb{S}^{k+1} \cong \mathbb{S}^k * \mathbb{S}^0,$$

which is easily seen to be homeomorphic to

$$\mathbb{S}^k \times [-1, +1] / \sim,$$

where $\sim$ is defined by $(x, t) \sim (x', t')$ if either $(x, t) = (x', t')$ or $t = t' \in \{\pm 1\}$. In other words, we obtain $\mathbb{S}^{k+1}$ by taking the cylinder $\mathbb{S}^k \times [-1, +1]$ and collapsing its ends, $\mathbb{S}^k \times \{-1\}$ and $\mathbb{S}^k \times \{+1\}$, each to a point. Obviously, the antipodal action on this model of the sphere is induced by the map $\nu(x, t) = (-x, -t)$ on the cylinder. It therefore suffices to construct a $\mathbb{Z}_2$-equivariant map

$$\mathbb{S}^k \times [-1, +1] \to |\operatorname{Hom}(K_2, M_r G)|$$

that is constant on each end $\mathbb{S}^k \times \{-1\}$ and $\mathbb{S}^k \times \{+1\}$.

Now consider the sequence of poset maps

$$\operatorname{Hom}(K_2, G) \times \operatorname{Hom}(K_2, P_r^0) \overset{\times}{\to} \operatorname{Hom}(K_2, G \times P_r^0) \overset{\pi}{\to} \operatorname{Hom}(K_2, M_r G),$$

where π is induced from the projection

$$G \times P_r^0 \to M_r G.$$

Note that this map is, by construction, equivariant with respect to the $\mathbb{Z}_2$-action given by α. Since we already have the map $f : \mathbb{S}^k \to |\operatorname{Hom}(K_2, G)|$, we are now going to investigate $\operatorname{Hom}(K_2, P_r^0)$. It turns out that $|\operatorname{Hom}(K_2, P_r^0)|$ is an interval with the multihomomorphism $i \mapsto \{0\}$ in its center. Figure 2.20 shows the partially ordered set $\operatorname{Hom}(K_2, P_3^0)$ and (in this 1-dimensional case also) the simplicial complex $\Delta(\operatorname{Hom}(K_2, P_r^0))$.

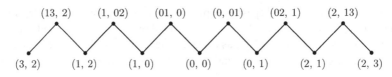

Fig. 2.20 Hom(K_2, P_3^0)

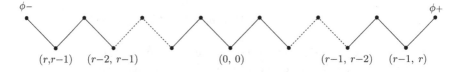

Fig. 2.21 The extension Q of Hom(K_2, P_r^0)

Note that the $\mathbb{Z}_2$-action α on $|\operatorname{Hom}(K_2, P_r^0)|$ is given by switching both ends of the interval. We now extend the poset Hom(K_2, P_r^0) to a poset Q by adding two new elements ϕ_+ and ϕ_- and extend the above poset map $\pi \circ \times$ to a map

$$\operatorname{Hom}(K_2, G) \times Q \xrightarrow{\varphi} \operatorname{Hom}(K_2, M_r G).$$

The extended order is defined by the cover relations $\phi_+ > (r-1, r)$ and $\phi_- > (r, r-1)$. In general, we obtain a picture as shown in Fig. 2.21.

We extend the action of α to Q by defining $\alpha * \phi_\pm = \phi_\mp$, and hence there exists an equivariant homeomorphism $g : [-1, +1] \rightarrow |Q|$ with respect to the $\mathbb{Z}_2$-action $t \mapsto -t$ on the interval.

Now the extension of the poset map $\pi \circ \times$ is defined by $\varphi((A, B), \phi_+) = (V(G) \times \{r-1\}, \{v^*\})$ and $\varphi((A, B), \phi_-) = \varphi(\alpha * ((B, A), \phi_+)) = \alpha * \varphi((B, A), \phi_+) = (\{v^*\}, V(G) \times \{r-1\})$. This obviously yields an equivariant poset map that is constant when restricted to either Hom$(K_2, G) \times \{\phi_+\}$ or Hom$(K_2, G) \times \{\phi_-\}$. Altogether, we obtain the following equivariant continuous map

$$\mathbb{S}^k \times [-1, 1] \xrightarrow{f \times g} |\operatorname{Hom}(K_2, G)| \times |Q| \xrightarrow{|\varphi|} |\operatorname{Hom}(K_2, M_r G)|,$$

which is constant on each end $\mathbb{S}^k \times \{-1\}$ and $\mathbb{S}^k \times \{+1\}$ as desired.   □

This lemma, together with Proposition 2.22, Lemma 2.21, and the previous observation, yields the following inequalities:

$$\chi(G) + 1 \geq \chi(M_r G) \geq \operatorname{co-ind}(|\operatorname{Hom}(K_2, M_r G)|) + 2$$

$$\geq \operatorname{co-ind}(|\operatorname{Hom}(K_2, G)|) + 3.$$

In the case that the topological lower bound on the chromatic number of G is tight, i.e., $\chi(G) = \operatorname{co-ind}(|\operatorname{Hom}(K_2, G)|) + 2$, we obtain

$$\chi(G) + 1 \geq \chi(M_r G) \geq \text{co-ind}(|\operatorname{Hom}(K_2, M_r G)|) + 2 \geq \chi(G) + 1.$$

In other words, the chromatic number increases by exactly one and the topological lower bound remains tight. By iterating this procedure we obtain the following.

Proposition 2.27. *Let G be a graph with $\chi(G) = \text{co-ind}(|\operatorname{Hom}(K_2, G)|) + 2$ and $r_1, \ldots, r_s \geq 2$. Then, for the iterated Mycielski construction $H = M_{r_1}(M_{r_2}(\cdots M_{r_s}(G) \cdots))$, we obtain*

$$\chi(H) = \text{co-ind}(|\operatorname{Hom}(K_2, H)|) + 2 = \chi(G) + s. \qquad \square$$

The immediate examples of graphs with tight topological lower bound are cycles and the family of Kneser graphs. Indeed, we have for even cycles

$$\chi(C_{2r}) = 2 = \text{co-ind}(|\operatorname{Hom}(K_2, C_{2r})|) + 2,$$

for odd cycles

$$\chi(C_{2r+1}) = 3 = \text{co-ind}(|\operatorname{Hom}(K_2, C_{2r+1})|) + 2,$$

and for the Kneser graphs

$$\chi(KG_{n,k}) = n - 2k + 2 = \text{co-ind}(|\operatorname{Hom}(K_2, KG_{n,k})|) + 2.$$

Exercises

1. Show that for any $n > d \geq 1$, there exists a set X of n vectors on the d-dimensional sphere $\mathbb{S}^d \subseteq \mathbb{R}^{d+1}$ such that any subset $S \subseteq X$ with $|S| = d + 1$ elements is linearly independent.

2. Show that the sets U_i defined in the proof of the Kneser conjecture by Greene on page 39 are open.

3. Prove the following lemma, known as Gale's lemma [Gal56]. For every $d \geq 0$ and $k \geq 1$, there exists a subset $X \subseteq \mathbb{S}^d$ of $2k + d$ points such that every open hemisphere contains at least k points of X, i.e., for all $x \in \mathbb{S}^d$, the intersection $X \cap \{y \in \mathbb{S}^d : \langle x, y \rangle > 0\}$ contains at least k elements. Hint: Consider the set of points $\{(-1)^i (1, i, i^2, \ldots, i^d) : i = 1, \ldots, 2k + d\}$.

4. Give a proof of Lovász's theorem along the lines of Greene's proof using only open sets. You will probably find a proof that was originally given by Bárány [Bár78]. Hint: Use Gale's lemma from the previous exercise.

5. Consider the following induced subgraph $SG_{n,k}$ of the Kneser graph $KG_{n,k}$ defined by Schrijver [Sch78]. The vertices are given by all k-subsets S of $[n]$ such that S does not contain a pair of consecutive numbers modulo n, i.e., none of the pairs $\{1, 2\}, \{2, 3\}, \ldots, \{n - 1, n\}, \{n, 1\}$ is contained is S. We call these sets *stable*. Since $SG_{n,k}$ is an induced subgraph of $KG_{n,k}$, any two stable sets are adjacent if and only if they are disjoint. Prove that $\chi(SG_{n,k}) = \chi(KG_{n,k}) = n - 2k + 2$. Hint: Use the ideas of the previous two exercises and the following interesting observation on polynomials.

Observation: Let $p(x)$ be a polynomial of degree at most d, with real coefficients. Then there exists a stable k-subset S of $[2k+d]$ such that $(-1)^i p(i) > 0$ whenever $i \in S$.

6. Show that the graphs $SG_{n,k}$ defined in the previous exercise are vertex critical with respect to the chromatic number, i.e., after removing an arbitrary vertex, the chromatic number drops by at least one. Remark: This is a hard exercise. At some point you might want to get some inspiration from Schrijver's article [Sch78].

7. Describe the neighborhood and Lovász complexes of $SG_{2n+1,n}$ and, furthermore, the action of ν on the Lovász complex.

8. Let G be a finite simple graph and assume that $\mathrm{ind}(\mathcal{L}(G)) \geq l + m + 2$. Show that G has a complete bipartite $K_{l,m}$ as a subgraph. This result is due to Csorba, Lange, Schurr, and Wassmer [CLSW04].

9. Show that there exists a homeomorphism from $\mathbb{S}^{n-2}$ to the subspace $S = \mathbb{S}^{n-1} \cap \{x : \sum_{i=1}^n x_i = 0\} \subseteq \mathbb{R}^n$ that is equivariant with respect to the antipodal actions as needed in the proof of Proposition 2.6 on page 46.

10. Show that the map φ defined in the proof of Proposition 2.6 on page 46 is bijective.

11. Show that there is a natural homeomorphism between $|\mathrm{Hom}(T,G)|$ and $|\Delta(\mathrm{Hom}(T,G))|$.

12. Show that the definition of composition $*$ of graph multihomomorphisms as given on page 55 is well defined, associative, and respects the inclusion order.

13. Fill in the details of the proof of Proposition 2.12.

14. Show that the $\mathbb{Z}_2$-action on $\mathrm{Int}(C(G))$ as defined on page 58 corresponds to the $\mathbb{Z}_2$-action on $\mathcal{L}(G)$ under the homeomorphism of the geometric realizations.

15. Show that in fact, $|\mathrm{Hom}(K_2, K_n)|$ is $\mathbb{Z}_2$-equivariant *homeomorphic* to the sphere $\mathbb{S}^{n-2}$ endowed with the antipodal action.

16. Show that, as needed in the proof of Lemma 2.23, there exists a simplicial approximation $\mathcal{H}$ of H such that $\mathcal{H}$ is odd on one of the boundary k-spheres and even on the other.

17. Let G be a finite graph with $\chi(G) = \mathrm{ind}(|\mathrm{Hom}(K_2, G)|) + 2$ and let $c : V(G) \to C$ be a proper coloring with $|C| = \chi(G)$. Show that for every partition $C = A \dot\cup B$ of the color set with $A, B \neq \emptyset$, there exists a complete bipartite subgraph $K_{|A|,|B|}$ of G such that one of the shores is colored with all of A, and the other with all of B. This result is due to Simonyi and Tardos [ST07].

18. Show by elementary means that the Mycielski graph $M_2 G$ has chromatic number $\chi(M_2 G) = \chi(G) + 1$.

Chapter 3
Evasiveness of Graph Properties

In many real-world situations we are forced to draw conclusions based only on partial information. For example, when we buy a used car it is infeasible to check every single part of the car. Yet an experienced person is able to almost guarantee the reliability of a car after only a certain relatively small number of checks.

In this chapter we investigate graph properties and whether it is possible to decide whether a given graph has a certain property based only on partial information about the graph. The exposition is to some extent based on [Aig88], [Bol04], [Schy06], and [KSS84].

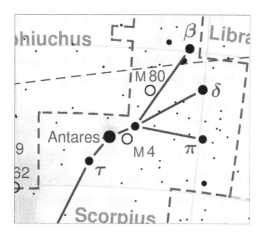

Fig. 3.1 A part of the Scorpius star constellation [Bro03]

3.1 Graph Properties and Their Complexity

In this chapter we consider graphs on a fixed set of n vertices, say $V = \{1, \ldots, n\}$. A simple graph $G = (V, E)$ is then determined by its edge set $E \subseteq \binom{V}{2}$, which allows us to identify G with E.

M. de Longueville, *A Course in Topological Combinatorics*, Universitext,
DOI 10.1007/978-1-4419-7910-0_3,
© Springer Science+Business Media New York 2013

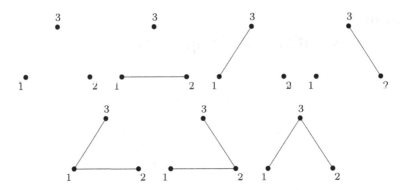

Fig. 3.2 The graphs on the vertex set $V = \{1, 2, 3\}$ with at most two edges

Graph Properties

We are interested in graph properties such as planarity, connectedness, and acyclicity. Graph properties are by definition required to be isomorphism invariant. In other words, if $\mathcal{P}$ is a property of graphs on the vertex set V, then a graph $G = (V, E)$ has property $\mathcal{P}$ if and only if any isomorphic copy $G' = (V, E')$ has property $\mathcal{P}$.

Since we consider graphs on a fixed vertex set V, each graph $G = (V, E)$ that we consider is determined by its edge set E. Hence, a property $\mathcal{P}$ of graphs on the vertex set V can be identified with the family of edge sets of graphs satisfying the property.

Definition 3.1. Let V be a fixed set of $n \geq 1$ vertices. A *graph property* $\mathcal{P}$ is a family of subsets of $\binom{V}{2}$ such that for any two isomorphic graphs $(V, E) \cong (V, E')$, either $E, E' \in \mathcal{P}$ or $E, E' \notin \mathcal{P}$.

As a first example, assume $n = 3$ and consider the property $\mathcal{P}_{3,2}$ of having at most two edges. Then

$$\mathcal{P}_{3,2} = \{\emptyset, \{12\}, \{13\}, \{23\}, \{12, 13\}, \{12, 23\}, \{13, 23\}\},$$

where we used the standard abbreviation uv for the edge $\{u, v\}$; cf. Fig. 3.2.

Hide and Seek

We consider a game for two players, let's call them Bob and Alice. They fix a vertex set V and a graph property $\mathcal{P}$ of graphs with vertex set V. The idea of the game is roughly that Bob imagines a graph and Alice wants to find out whether Bob's graph has property $\mathcal{P}$.

From the viewpoint of Alice, Bob may have a fixed graph in mind. But Bob can change which graph he is imagining after each of Alice's questions, as long as it is consistent with the information he has already given.

A game takes place as follows. Alice asks questions of the type, "Is e an edge of the graph?" for potential edges $e \in \binom{V}{2}$, and Bob answers in each case with *yes* or *no*, thereby revealing information about the graph's edges and nonedges. So in each stage of the game Alice has partial information about Bob's graph: according to his answers, she knows about some edges that are in the graph and some that are not. Let's call these sets of edges Y and N. She wants to decide as quickly as possible whether Bob's graph has property $\mathcal{P}$. But what does that mean? Let's call any graph $G = (V, E)$ on the vertex set V a *completion of the partial graph defined by* (Y, N) if $Y \subseteq E$ and $E \cap N = \emptyset$. So the graph that Bob has in mind is such a completion. Alice wants to decide as quickly as possible whether every such completion of the partial graph defined by (Y, N) has property $\mathcal{P}$ or whether every completion of the partial graph does not have property $\mathcal{P}$. Bob, on the other hand, wants Alice to ask as many questions as possible.

For the graph property $\mathcal{P}_{3,2}$ given above, Alice might ask as follows: "Is 12 an edge of the graph?" If Bob's answer is *no*, any completion of the graph has at most two edges, and Alice can answer, "The graph has property $\mathcal{P}_{3,2}$!" If the answer is *yes*, she must keep asking. Possibly, "Is 13 an edge of the graph?" If the answer is *no*, Alice is done; if it is *yes*, she indeed has to ask the third question, "Is 23 an edge of the graph?" We see that if Bob always answers the first two questions with *yes*, then Alice cannot do better than to ask all $\binom{3}{2} = 3$ potential questions.

As noted above, Bob does not necessarily have to have a fixed graph in mind. He may decide after each of Alice's question which answer suits his goal best. But from Alice's viewpoint, Bob may already have a certain fixed graph in mind. We can therefore also say that Alice wants to decide with certainty, and as quickly as possible, whether this *hypothetical graph* has property $\mathcal{P}$. At a particular stage of the game, this hypothetical graph may be just any completion of the partial graph that Alice has knowledge about so far.

Strategies

A *strategy* ϕ *of the seeker Alice* is an algorithm that, depending on Bob's answers at each stage of the game, either assigns an edge that Alice uses for her next question or if possible gives one of the following answers: "The graph has property $\mathcal{P}$!" or "The graph does not have property $\mathcal{P}$!"

Alice's strategy discussed in our previous example is shown schematically in Fig. 3.3. For obvious reasons, such an algorithm is called a *decision-tree algorithm*.

A *strategy* ψ *of the hider Bob* is given by a map that assigns to each triple (Y, N, e) one of the answers *yes* or *no*, where $Y, N \subseteq \binom{V}{2}$ are disjoint edge sets and $e \in \binom{V}{2} \setminus (Y \cup N)$ is an edge in the complement. The sets Y and N represent the sets of edges that Bob previously has answered with *yes*, respectively *no*, and e is the edge about which Alice is currently asking. The pairs (Y, N) are called the *stages of the game,* and their evolution completely describes the course of the game.

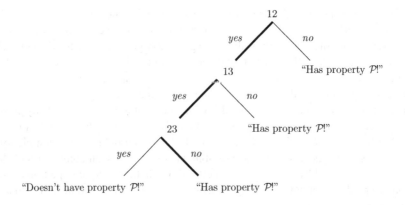

Fig. 3.3 The strategy of the seeker

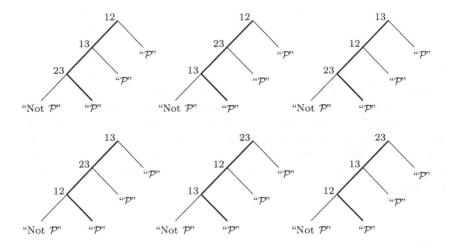

Fig. 3.4 All possible strategies of the seeker and one particular strategy of the hider

For the example $\mathcal{P}_{3,2}$, a strategy for Bob is given by the map

$$(Y, N, e) \longmapsto \begin{cases} yes, & \text{if } |Y \cup \{e\}| = |Y| + 1 \leq 2, \\ no, & \text{otherwise.} \end{cases}$$

A strategy of Bob determines a path from the top of the tree to a leaf in each decision-tree corresponding to a strategy of Alice. Figure 3.4 shows all possible strategies of Alice and the paths determined—drawn in bold—by Bob's strategy that we just defined. We observe that with this particular strategy of Bob's, Alice is always forced to ask the maximal number $\binom{3}{2}$ of questions.

Complexity

Alice wants to play with a fixed strategy that is optimal for her, i.e., a strategy that minimizes the maximal length of a game with respect to all of Bob's strategies. The complexity of a graph property is a measure for precisely this length.

Definition 3.2. The *complexity* $c(\mathcal{P})$ *of a graph property* $\mathcal{P}$ is the minimal number k for which there exists a seeker's strategy such that regardless of the hider's strategy, the seeker needs to ask at most k questions.

For our simple example above, the complexity is $c(\mathcal{P}) = \binom{3}{2} = 3$, as discussed. In order to phrase this definition in a formula, let $c(\mathcal{P}, \phi, \psi)$ be the number of questions that Alice has to ask when she is playing with strategy ϕ and Bob is playing with ψ. Then by definition,

$$c(\mathcal{P}) = \min_{\phi} \max_{\psi} c(\mathcal{P}, \phi, \psi),$$

where ϕ and ψ run through all possible strategies of the seeker and hider. Since a simple graph on n vertices has at most $\binom{n}{2}$ edges, we clearly have $c(\mathcal{P}) \leq \binom{n}{2}$. We call a strategy ϕ_0 for Alice *optimal* if it attains the minimum, i.e.,

$$c(\mathcal{P}) = \max_{\psi} c(\mathcal{P}, \phi_0, \psi).$$

With this new language in hand, we want to discuss the idea of a *hypothetical graph* in Bob's mind mentioned earlier. If $G = (V, E)$ is an arbitrary graph, then G defines a particular strategy ψ_G for Bob. Namely,

$$(Y, N, e) \longmapsto \begin{cases} yes, & \text{if } e \in E, \\ no, & \text{if } e \notin E. \end{cases}$$

In other words, Bob has chosen the graph G and answers according to its edges and nonedges.

Lemma 3.3. *For any graph property $\mathcal{P}$ of graphs on the vertex set V,*

$$c(\mathcal{P}) = \min_{\phi} \max_{\psi} c(\mathcal{P}, \phi, \psi) = \min_{\phi} \max_{G} c(\mathcal{P}, \phi, \psi_G),$$

where the right-side maximum is taken over all possible graphs G with vertex set V.

Proof. Let ϕ and ψ be arbitrary strategies for Alice and Bob. It suffices to show that there exists a graph G such that $c(\mathcal{P}, \phi, \psi) = c(\mathcal{P}, \phi, \psi_G)$. But this is easy. If Alice and Bob play according to the strategies ϕ and ψ, and if Y is the set of edges that Bob has answered during the game with *yes*, then let G be the graph $G = (V, Y)$. □

In other words, for Alice to choose an optimal strategy it does not matter whether Bob is playing with a fixed graph in mind or is constructing the graph during the game.

Evasiveness

Let's consider some extreme cases for the values of $c(\mathcal{P})$. If the graph property is empty, $\mathcal{P} = \emptyset$, in other words no graph has property $\mathcal{P}$, then the seeker Alice can answer right away: "The graph does not have property $\mathcal{P}$!" Similarly, if all graphs satisfy property $\mathcal{P}$, i.e., $\mathcal{P}$ is the set of all subsets of $\binom{V}{2}$, Alice can answer immediately. We call these two properties the *trivial graph properties*, and in these cases the complexity is zero: the seeker Alice does not need to ask a single question. Note that there are no other cases of graph properties with complexity zero.

The other extreme is more interesting, namely the case in which the complexity is $\binom{n}{2}$, i.e., the maximal number. An easy class of examples with this complexity is given by a generalization of our introductory example. For fixed $n \geq 2$ and $0 \leq k < \binom{n}{2}$, let $V = \{1, \dots, n\}$ and consider the graph property

$$\mathcal{P}_{n,k} = \left\{ E \subset \binom{V}{2} : |E| \leq k \right\},$$

i.e., all graphs on the vertex set V with at most k edges. For this property a possible strategy for the hider Bob might be to answer the first k questions with *yes*, and all others with *no*. In other words, regardless of Alice's strategy, she knows already after the first k questions about the existence of k edges in the graph. But then she has to keep asking about all other edges to make sure that there are not more than k edges in the graph. Hence the complexity is $c(\mathcal{P}_{n,k}) = \binom{n}{2}$, the maximal possible number.

We call all properties $\mathcal{P}$ with maximal complexity $c(\mathcal{P}) = \binom{n}{2}$ *evasive,* as they "tend to avoid self-revelation" [JA01].

Most nontrivial graph properties turn out to be evasive. We will see quite a few examples later in this chapter.

The Greedy Strategy

One particular strategy for the hider Bob suggests itself, the following *greedy strategy*. Bob answers *yes* whenever the graph constructed so far is contained in a graph with property $\mathcal{P}$, and *no* otherwise. More precisely, consider a particular step in the game when the seeker asks, "Is e an edge of the graph?" and by the previous answers knows the existence of a set Y of edges and a set N of nonedges already. The greedy strategy yields the answer *yes* whenever there exists an edge set $E \in \mathcal{P}$ disjoint from N such that $Y \cup \{e\} \subseteq E$, and *no* otherwise.

In our previous example for the property *having at most k edges*, the strategy we described is the greedy strategy.

Lemma 3.4. *Assume that Bob is playing the greedy strategy and that the game is in stage (Y, N), i.e., Alice has knowledge about the existence of a set Y of edges and a set N of nonedges. Then any F with $Y \subseteq F \in \mathcal{P}$ is disjoint from N.*

Proof. Assume that there exists an $F \in \mathcal{P}$ such that $F \cap N \neq \emptyset$. Let $e \in F \cap N$ be the edge that came first in the order of Alice's questions. Then clearly because of the existence of $F \in \mathcal{P}$, Bob would have had to answer *yes* to Alice's question, "Is e an edge of the graph?" A contradiction.   □

The following lemma tells us when the greedy strategy witnesses evasiveness of the graph property.

Lemma 3.5. *Let $\mathcal{P} \neq \emptyset$ be a graph property, ϕ any strategy for the seeker Alice, and ψ the greedy strategy for the hider Bob. Then*

$$c(\mathcal{P}, \phi, \psi) = \binom{n}{2}$$

if for each $E \in \mathcal{P}$ and $e \in E$ with $E \setminus \{e\} \in \mathcal{P}$, there exist an $f \in \binom{V}{2} \setminus E$ and $F \in \mathcal{P}$ such that $(E \setminus \{e\}) \cup \{f\} \subseteq F$.

Proof. Assume to the contrary that Alice has not yet asked $\binom{n}{2}$ questions and can already decide in stage (Y, N), "The graph has property $\mathcal{P}$!" Then $Y \cup N \neq \binom{V}{2}$ and each E with $Y \subseteq E \subseteq \binom{V}{2} \setminus N$ satisfies $E \in \mathcal{P}$. Set $E = \binom{V}{2} \setminus N$ and choose an arbitrary

$$e \in E \setminus Y = \left(\binom{V}{2} \setminus N \right) \setminus Y = \binom{V}{2} \setminus (Y \cup N) \neq \emptyset.$$

Then $E \setminus \{e\} \in \mathcal{P}$, and by assumption there exist $f \in \binom{V}{2} \setminus E = N$ and an $F \in \mathcal{P}$ such that $(E \setminus \{e\}) \cup \{f\} \subseteq F$. In particular, $f \in F \cap N$, in contradiction to the previous lemma.   □

This simple criterion proves that quite a few graph properties are evasive.

Theorem 3.6. *Let $n \geq 3$ and $0 \leq k < \binom{n}{2}$. The following properties $\mathcal{P}$ of graphs on n vertices are evasive:*

1. *The graph has at most k edges.*
2. *The graph has exactly k edges.*
3. *The graph is acyclic, i.e., it does not contain cycles.*
4. *The graph is a spanning tree.*
5. *The graph is connected.*

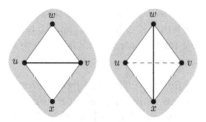

Fig. 3.5 The edge $e = uv$ and its neighboring triangles

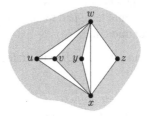

Fig. 3.6 The graph G when wx is present

Proof. In all of these cases, the condition of Lemma 3.5 is easily checked.   □

As an interesting application of Lemma 3.5, we show that planarity is an evasive property.

Theorem 3.7. *For $n \geq 5$, the property of being planar is evasive.*

Proof. Let E be the edge set of a planar graph and $e = uv \in E$. Since $E \setminus \{e\}$ is always a planar graph, we have to show that there is an edge $f \in \binom{V}{2} \setminus E$ such that $(E \setminus \{e\}) \cup \{f\}$ is planar. We may assume that E is the edge set of a maximal planar graph, since the statement is obvious otherwise. Consider a planar drawing of the graph $G = (V, E)$. All faces of this drawing are triangles, and the edge e is an edge of two neighboring triangular faces, say uvw and uvx. If $wx \notin E$, then $f = wx$ satisfies our needs, since after removal of e, we can draw the diagonal connecting w and x. Compare Fig. 3.5. The gray regions in the figure depict the unknown rest of the graph.

But if wx is an edge in G, then let wxy and wxz be the two triangles neighboring wx, as illustrated in Fig. 3.6.

There are two cases to consider. The first case is that the pair of vertices $\{u, v\}$ and $\{y, z\}$ are identical. Then the graph is the complete graph K_4 as shown in Fig. 3.7, which contradicts our assumption $n \geq 5$.

In the other case, the two pairs $\{u, v\}$ and $\{y, z\}$ are not identical. Let us assume that $u \notin \{y, z\}$, since the situation in which $v \notin \{y, z\}$ works analogously. Then yz cannot be an edge of G because in the drawing it would have to intersect one of the edges wx, uw, ux. This follows from the Jordan curve theorem, Theorem A.9. See Fig. 3.8.

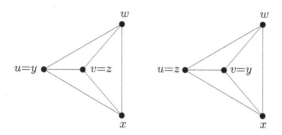

Fig. 3.7 The two cases in which $\{u, v\}$ and $\{y, z\}$ are identical

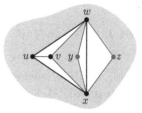

Fig. 3.8 The vertices y and z are separated by the circle $uwxu$

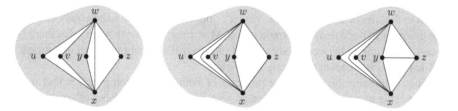

Fig. 3.9 Redrawing the edge wx and introducing yz in the drawing of $G \setminus \{uv\}$

Hence, in $G \setminus \{uv\}$ we may redraw the edge wx in the quadrilateral that appears after removal of uv, and draw the edge yz afterward. The whole procedure is shown in Fig. 3.9. In other words, $f = yz$ satisfies the required needs.   □

Nonevasive Graph Properties

There are only a few graph properties known to be not evasive. For an easy start, we will describe a nonevasive graph property of graphs on a set V of six vertices. Let $\mathcal{B}_6$ be the graph property given by all possible edge sets E of graphs $G = (V, E)$ that are isomorphic to one of the three graphs shown in Fig. 3.10. It is an easy exercise to see that property $\mathcal{B}_6$ is nonevasive.

Fig. 3.10 The three isomorphism classes of a nonevasive graph property

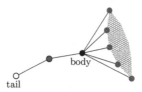

Fig. 3.11 A scorpion graph

We will now describe a whole family of nonevasive graph properties due to Best, van Emde Boas, and Lenstra [BEBL74]. Let $n \geq 5$ be fixed. A graph on n vertices is called a *scorpion graph* if it contains two nonadjacent vertices of degree 1 and degree $n - 2$, called *tail* and *body*, whose uniquely defined common neighbor (playing the role of Antares; cf. Fig. 3.1) has degree 2. There are no restrictions on the adjacencies among the remaining $n - 3$ vertices, see Fig. 3.11.

Note that the body is the unique vertex of degree $n - 2$, and hence body and tail are uniquely defined in any scorpion graph. Moreover, it is worth noting that there is a certain symmetry between body and tail: the body is adjacent to all but one vertex, while the tail is adjacent to exactly one vertex.

The remarkable fact about the property of being a scorpion graph is that it is recognizable in a linear number of steps with respect to the number of vertices. In particular, the property of being a scorpion graph is nonevasive for large enough n.

Theorem 3.8 (Best, van Emde Boas, and Lenstra [BEBL74]). *Let $\mathcal{P}$ be the property of being a scorpion graph on $n \geq 5$ vertices. Then the complexity of $\mathcal{P}$ is bounded by $c(\mathcal{P}) \leq 6n - 13$.*

Proof. We play the role of the seeker Alice and describe an algorithm that determines in the required number of steps whether the hypothetical graph of the hider Bob is a scorpion graph.

The idea is to determine a unique single candidate for a body or a tail vertex. During the course of the game the vertices will be categorized into body and tail candidates. We will refer to the edges to which Bob answers with *yes* as accepted edges and to the edges with answer *no* as rejected edges. Note that a body candidate is a vertex that has at most one rejected incident edge, and a tail candidate is a vertex with at most one accepted incident edge. The algorithm has three phases. The first phase serves to *partition* the vertex set into body and tail candidates. The second phase reduces at least one of these sets to at most one candidate. With a unique body or tail candidate the third, and final, phase decides whether the hypothetical graph is a scorpion graph. After the description of the three phases, we count the questions that are needed.

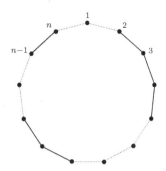

Fig. 3.12 The knowledge of the seeker after the first n questions

Phase I: Each vertex of $V = \{1, \ldots, n\}$ can uniquely be classified

- As *body candidate* if it has at most one rejected and at least two accepted incident edges, and
- As *tail candidate* if it has at most one accepted and at least two rejected incident edges.

In order to classify the vertices, we have to check at most three incident edges per vertex. We start with the edge set of the cycle $1, 2, \ldots, n, 1$, i.e., the edges $12, 23, 34, \ldots, (n-1)n, n1$. After these n questions, we know about two incident edges of each vertex, and the result might look like Fig. 3.12. Accepted edges are shown in bold; rejected edges are dotted.

The vertices with two accepted incident edges (such as vertex 3 in the figure) already qualify as body candidates, while the vertices with two rejected incident edges (such as vertex 1 in the figure) qualify as tail candidates. The remaining vertices (e.g., vertex 2 in the figure) are still indifferent; they may qualify as body or tail vertices. We denote the set of indifferent vertices by I. If I is empty, then either all edges have been rejected, or all edges have been accepted. In the former case, none of the vertices can be a body, while the latter case implies that none of the vertices can be a tail. Either way, we can already answer, "The graph is not a scorpion graph!" Otherwise, this set I has even cardinality greater than or equal to 2. The case $|I| = 2$, i.e., I consists of two adjacent vertices only, is an easy exercise that we leave to the reader. In all other cases, $|I| \geq 4$, and we may divide I into pairs of vertices that are not adjacent on the cycle. We are now asking for exactly those edges that are given by these pairs. The result might look like Fig. 3.13. Now each vertex is either a body or a tail candidate, which we depict in the figure by a black, resp. white, bullet.

Phase II: We want to single out a unique candidate for either the body or the tail. In order to do so, we will start by assigning a weight of 1 or 2 to each vertex. We will then successively ask for edges and, after each step, adjust weights in such a way that exactly one weight reduces by one and all others remain fixed. If a vertex

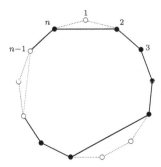

Fig. 3.13 The knowledge of the seeker after $n + \frac{|I|}{2}$ questions at the end of Phase I

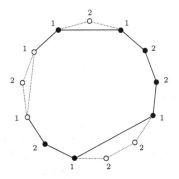

Fig. 3.14 The initial weights after Phase I

obtains weight 0 in this way, it is no longer considered a candidate, and we will no longer ask for edges incident to this vertex.

The weights are defined differently for body and tail candidates as follows:

- A body candidate obtains weight 2 minus the number of rejected incident edges, while
- a tail candidate obtains weight 2 minus the number of accepted incident edges.

Figure 3.14 shows the initial weights after Phase I.

Initially, the *total weight,* i.e., the sum of all weights, is obviously equal to $2n - |I|$. We now successively ask for edges with one vertex in the set of body candidates, and one vertex in the set of tail candidates. Say v is a body candidate and w is a tail candidate. We ask, "Is vw an edge of the graph?" If the answer is *yes,* then the number of accepted edges incident to w increases by 1, and hence the weight of w decreases by 1. Similarly, if the answer is *no,* the number of rejected edges incident to v increases, and hence the weight of v decreases by 1. If either of the two weights drops to zero, the vertex of weight zero is no longer considered a candidate, since it has at least two accepted and two rejected incident edges. A few iterations of this procedure are shown in Fig. 3.15.

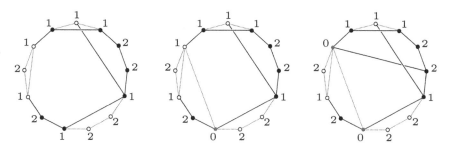

Fig. 3.15 A few iterations of Phase II

Since the total weight decreases by one in each iteration, after at most $2n - |I| - 2$ steps one of the three following situations occurs:

1. The set of body or tail candidates becomes empty.
2. The sets of body and tail candidates are both nonempty and one of them contains exactly one element.
3. There are sets of $b \geq 1$ body candidates and $t \geq 1$ tail candidates remaining, where $\min\{b, t\} \geq 2$, and all edges between these sets have already been asked for.

In situation (a), we can obviously answer, "The graph is not a scorpion graph!" In situation (b), we proceed to Phase III. Now assume we are in situation (c). First of all, note that the total weight is at least three. Therefore at most $2n - |I| - 3$ questions have been asked so far in Phase II. Now let e be the number of accepted edges between the body and tail candidates. Since each body candidate must be adjacent to all but at most one tail candidate, we have $b(t - 1) \leq e$. Similarly, since each tail candidate has at most one neighbor among the body candidates, we have $e \leq t$. Hence $(b - 1)(t - 1) \leq 1$, and since $\min\{b, t\} \geq 2$, we must have $b = t = 2$. Let's say the body candidate vertices are $\{a, b\}$ and the tail candidate vertices are $\{c, d\}$. Then $2 = b(t - 1) \leq e \leq t = 2$, and hence there are exactly two edges between the body and tail candidates. Without loss of generality, the edges are ac and bd. Then there are only two possibilities for a scorpion graph: either a is the body and d the tail, or b is the body and c the tail. Compare Fig. 3.16. Let x be any vertex other than a, b, c, and d. Now we ask for the edge ax. If ax is an edge of the graph, then only a remains as body candidate. Conversely, if ax is not an edge, then only b remains as body candidate. We then proceed to Phase III.

Phase III: We are left with a unique body or tail candidate u. Assume u is a body candidate. We ask for the adjacency relations of u that are not yet known. There are at most $n - 3$ of them. In case the degree of u is $n - 2$, we now know the unique tail candidate. Checking its adjacency relations and the adjacency relations of the unique common neighbor requires at most another $n - 3 + n - 3$ questions. The case that u is a tail candidate is similar.

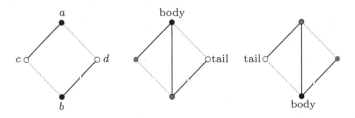

Fig. 3.16 The case $b = t = 2$

Step count: The three phases require at most $n + \frac{|I|}{2}$, $2n - |I| - 2$, and $3n - 9$
questions. Since $|I| \geq 4$, we obtain in total at most

$$n + \frac{|I|}{2} + 2n - |I| - 2 + 3n - 9 = 6n - \frac{|I|}{2} - 11 \leq 6n - 13. \qquad \square$$

3.2 Evasiveness of Monotone Graph Properties

In the previous section we have seen quite a variety of evasive graph properties and
a few nonevasive graph properties. A noticeable difference is that all nonevasive
graph properties share one common feature: they are not monotone, i.e., they are
not closed under removing (or alternatively adding) edges.

Monotone Graph Properties

Definition 3.9. A graph property $\mathcal{P}$ is called *monotone* if it is closed under
removing edges, i.e., whenever $E' \subseteq E \in \mathcal{P}$, then $E' \in \mathcal{P}$.

Obvious examples of monotone properties are the properties *having at most k edges*,
being planar, *being acyclic*, etc.
     Note that if a graph property $\mathcal{P}$ is closed under adding edges, then the *comple-
mentary property*

$$\overline{\mathcal{P}} = \left\{ \binom{V}{2} \setminus E : E \in \mathcal{P} \right\}$$

is monotone. The fact that $c(\overline{\mathcal{P}}) = c(\mathcal{P})$, as shown in Exercise 7, now justifies that
we concentrate only on properties closed under removing edges.

All known monotone graph properties besides the trivial ones—as defined on page 74—are evasive. This led Richard Karp in the early 1970s to the following conjecture.

Conjecture 3.10 (R. Karp).   Every nontrivial monotone graph property is evasive.

The main result of this section will show that the conjecture is true whenever $\mathcal{P}$ is a monotone property of graphs on n vertices and n is a power of a prime number. This was shown by topological methods in a striking paper [KSS84] by Jeff Kahn, Michael Saks, and Dean Sturtevant in 1984. In the same publication, they also proved the conjecture in the case $n = 6$, the smallest non-prime-power case. All other cases are still open even though a great deal of research has been carried out since then.

The proof of the prime-power case involves several steps. The first important step will be to link the property of evasiveness to a topological property. Subsequently, we will apply a somewhat deeper topological result along with a little bit of algebra. The topological background will be explained in more detail in Appendix E.

Simplicial Complexes

The monotonicity condition yields a direct link to topology: every monotone graph property defines an (abstract) simplicial complex.

In fact, consider a graph property $\mathcal{P}$ of graphs with vertex set V. Then $\mathcal{P}$ is an abstract simplicial complex on the vertex set $X = \binom{V}{2}$, i.e., a simplicial complex on the edge set of the complete graph on V. Each $E \in \mathcal{P}$ constitutes a face $E \subseteq X$ of the simplicial complex.

As a first example, we return to the property *having at most two edges* for 3-vertex graphs

$$\mathcal{P} = \{\emptyset, \{12\}, \{13\}, \{23\}, \{12, 13\}, \{12, 23\}, \{13, 23\}\},$$

with our usual abbreviation uv for the edge $\{u, v\}$.

The graph property $\mathcal{P}$ corresponds to the boundary of a 2-simplex with vertex set $X = \{12, 13, 23\}$ as shown in Fig. 3.17. The graph on V with no edges corresponds to the empty face, the graphs with precisely one edge correspond to the 0-dimensional faces of the complex, and the graphs with precisely two edges correspond to the 1-dimensional faces. Compare also Fig. 3.2 on page 70, where all graphs with property $\mathcal{P}$ are shown.

In general, the property *having at most k edges* in an n-vertex graph corresponds to the $(k - 1)$-skeleton of an $\left(\binom{n}{2} - 1\right)$-simplex.

The topological spaces associated with these examples are not completely trivial: they have "holes." It will turn out that any graph property is evasive as soon as the associated space is topologically nontrivial in a very strong sense. Before we make this precise by introducing the concept of *collapsibility*, we will first generalize our notion of evasiveness to general set systems and thereby to simplicial complexes.

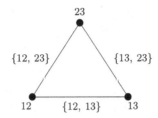

Fig. 3.17 The simplicial complex corresponding to the property *having at most two edges* in a 3-vertex graph

Evasiveness of Set Systems

We have seen that the graphs satisfying a given property correspond to the faces of the associated simplicial complex. We will now generalize the game between the hider and the seeker to arbitrary set systems.

The hider and seeker agree on a vertex set X and a set system $S \subseteq \mathcal{P}(X)$. Now the game is to decide whether a hypothetical subset $\sigma \subseteq X$—unknown to the seeker Alice—of the vertex set is an element of the set system S.

Alice follows a decision-tree algorithm with questions of the type " Is $x \in \sigma$?" for vertices $x \in X$, and the hider Bob answers *yes* or *no*. Alice's goal is to ask as few questions as possible, whereas the aim of Bob is to force Alice to ask as many questions as possible. The game is over as soon as Alice can decide whether $\sigma \in S$.

As before, the *complexity*, $c(S, X)$, is defined as the minimal number k such that there exists a strategy for Alice that allows her always to finish the game by asking at most k questions.

Definition 3.11. Let X be a set of m vertices and S a set system $S \subseteq \mathcal{P}(X)$. The pair (S, X) is called *evasive* if the complexity $c(S, X)$ is equal to m, i.e., for every strategy of the seeker Alice, there exists a subset $\sigma \subseteq X$ such that she needs to ask m questions in order to decide whether $\sigma \in S$.

We will be mostly interested in the case that the set system is a simplicial complex $K \subseteq \mathcal{P}(X)$.

There are two interesting observations to discuss. First of all, not all elements from X have to appear as vertices of K. This will be of some importance in the sequel.

Secondly, this type of game is more general than the game for graphs: the size of the vertex set of the simplicial complex can be arbitrary, while the number of edges of the complete graphs are always binomial coefficients. Moreover, a graph property is by definition invariant under graph isomorphism—which yields a certain symmetry of the associated simplicial complex that we will discuss later—whereas there is no such condition on the simplicial complexes we are now considering.

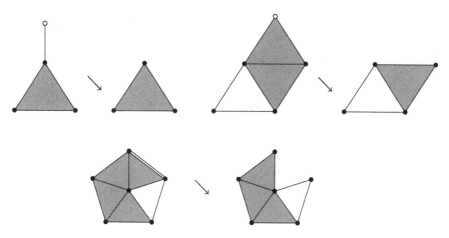

Fig. 3.18 Examples of elementary collapses

Furthermore, note that if $\mathcal{P}$ is a monotone graph property of graphs on the vertex set V, and $X = \binom{V}{2}$, then $\mathcal{P}$ is an evasive graph property if and only if $(\mathcal{P}, X)$ is evasive.

Collapsibility

Collapsibility is a property of simplicial complexes. Loosely speaking, a simplicial complex is collapsible if it can be deformed to a single vertex by a sequence of "scrunching steps." The scrunching steps are given by so-called elementary collapses. In order to define these, we first need to introduce the concept of a free face.

Definition 3.12. A nonempty face σ in a simplicial complex is a *free face* if

- It is not inclusion maximal in K, and
- It is contained in exactly one inclusion-maximal face of K.

An *elementary collapse* of K is a simplicial complex K' obtained from K by the removal of a free face $\sigma \in K$ along with all faces that contain σ, i.e., $K' = K \setminus \{\tau : \tau \in K, \sigma \subseteq \tau\}$. Whenever a complex K' is obtained from K by an elementary collapse, we denote this by $K \searrow K'$.

Figure 3.18 gives a few examples of elementary collapses. It is an easy exercise to show that an elementary collapse induces a homotopy equivalence of the polyhedra associated with K and K'. Similar to the concept of contractability, we introduce the concept of collapsibility.

Definition 3.13. A simplicial complex K is *collapsible* if there exists a sequence of elementary collapses $K = K_0 \searrow K_1 \searrow K_2 \searrow \cdots \searrow K_r = \{\emptyset, \{z\}\}$ onto a single vertex.

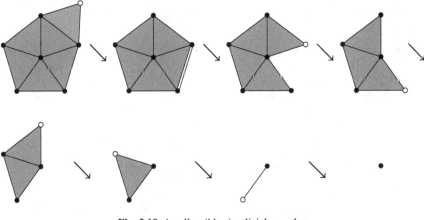

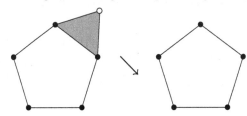

Fig. 3.19 A collapsible simplicial complex

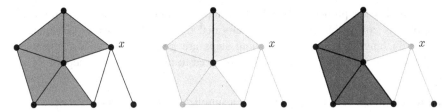

Fig. 3.20 A noncollapsible simplicial complex

Fig. 3.21 A complex and the link and deletion of the vertex x

Figure 3.19 shows a collapsible simplicial complex along with the sequence of elementary collapses, while Fig. 3.20 illustrates a simplicial complex that is not collapsible: it allows one elementary collapse, but afterward does not possess any more free faces.

Since elementary collapses induce homotopy equivalences, any collapsible complex is also contractible. (For simplicity, we will call a simplicial complex contractible if its polyhedron is contractible.) But note that the concepts of contractability and collapsibility are not identical. There exist contractible complexes that are not collapsible. A prominent example for such a complex is the dunce hat [Zee63] and Bing's house with two rooms [Bin64].

Link, Deletion, and Collapsibility

In order to show the main result of this section—as stated in the section title—we need the concept of two particular complexes that occur in our setting. Let $K \subseteq \mathcal{P}(X)$ be a simplicial complex and $x \in X$. The simplicial complexes *link* and *deletion* of x are defined to be

$$\mathrm{lk}(x, K, X) = \{\sigma \subseteq X \setminus \{x\} : \sigma \cup \{x\} \in K\},$$

$$\mathrm{del}(x, K, X) = \{\sigma \subseteq X \setminus \{x\} : \sigma \in K\}.$$

If no confusion about K and X can occur, we will abbreviate the two complexes by $\mathrm{lk}(x)$ and $\mathrm{del}(x)$. Note that if $x \in X \setminus \mathrm{vert}(K)$, then $\mathrm{lk}(x) = \emptyset$ and $\mathrm{del}(x) = K$. Also note that if x is an isolated vertex of K, i.e., when $\{x\}$ is a maximal face of K, then $\mathrm{lk}(x) = \{\emptyset\}$. A more illuminating example, where $x \in \mathrm{vert}(K)$, is given in Fig. 3.21.

Lemma 3.14. *Let* $K \subseteq \mathcal{P}(X)$ *be a simplicial complex and* $x \in X$. *If* $\mathrm{lk}(x)$ *and* $\mathrm{del}(x)$ *are collapsible, then so is* K.

Proof. Note that the collapsibility of $\mathrm{lk}(x)$ implies in particular that $\mathrm{lk}(x)$ is nonempty and hence $x \in \mathrm{vert}(K)$. It clearly suffices to show that K can be collapsed down to $\mathrm{del}(x)$. Since this is fairly straightforward, the reader is encouraged to provide the details accompanying the *picture proof* [Pól56] as shown in Figs. 3.22–3.24.   □

Nonevasive Complexes Are Collapsible

The following theorem establishes the essential link between evasiveness and topology.

Theorem 3.15. *Let* $X \neq \emptyset$ *and* $K \subseteq \mathcal{P}(X)$ *be a nonempty simplicial complex. If* (K, X) *is nonevasive, then* K *is collapsible.*

Proof. First of all, note that $K \neq \{\emptyset\}$, since $(\{\emptyset\}, X)$ is easily seen to be evasive for $X \neq \emptyset$. And hence K has at least one vertex. We now proceed by induction on $n = |X|$. The case $n = 1$ is clear. For the induction step $n \geq 2$ consider the decision-tree of an algorithm that proves nonevasiveness of (K, X) and assume "$x \in \sigma$?" is the first question according to the algorithm.

Denote by ϕ_L, respectively ϕ_R, the strategies belonging to the left branch L succeeding the *yes* answer to the first question, respectively the right branch R succeeding the *no* answer, as shown in Fig. 3.25.

Consider $\mathrm{lk}(x) = \mathrm{lk}(x, K, X)$ and $\mathrm{del}(x) = \mathrm{del}(x, K, X)$. We claim that $(\mathrm{lk}(x), X \setminus \{x\})$ and $(\mathrm{del}(x), X \setminus \{x\})$ are nonevasive. In order to see this, observe that for any subset $\tau \subseteq X \setminus \{x\}$, we have $\tau \in \mathrm{lk}(x)$ if and only if $\tau \cup \{x\} \in K$, and

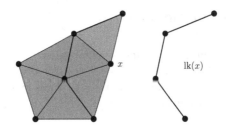

Fig. 3.22 The complex K, a vertex x, and $\mathrm{lk}(x)$

Fig. 3.23 The collapsing sequence of $\mathrm{lk}(x)$

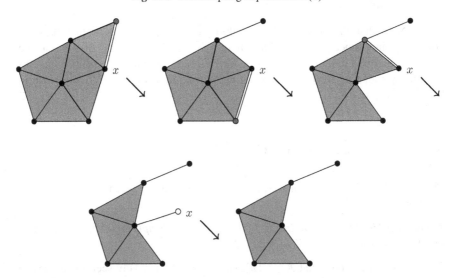

Fig. 3.24 Collapsing K onto $\mathrm{del}(X)$

$\tau \in \mathrm{del}(x)$ if and only if $\tau \in K$. Hence ϕ_L is a strategy proving nonevasiveness of $(\mathrm{lk}(x), X \setminus \{x\})$, and ϕ_R is a strategy proving nonevasiveness of $(\mathrm{del}(x), X \setminus \{x\})$.

We want to apply the induction hypothesis. We have to be a little careful about the possibility that $\mathrm{lk}(x)$ or $\mathrm{del}(x)$ may be empty.

If $\mathrm{lk}(x) = \emptyset$, then clearly $K = \mathrm{del}(x)$, which is nonempty by assumption. Since $(\mathrm{del}(x), X \setminus \{x\})$ is nonevasive, we obtain by the induction hypothesis that K is collapsible. If $\mathrm{del}(x) = \emptyset$, then by the fact that K has at least one vertex, we must

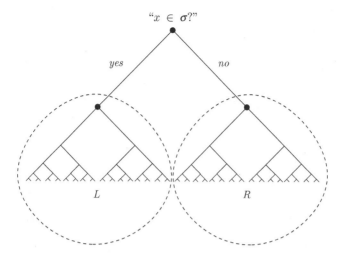

Fig. 3.25 The two branches L and R succeeding the first question of the seeker

have $K = \{\emptyset, \{x\}\}$, which is collapsible by definition. If neither $\mathrm{lk}(x)$ nor $\mathrm{del}(x)$ is empty, then by the induction hypothesis, both complexes are collapsible, and we are done by an application of Lemma 3.14.   □

3.3   Karp's Conjecture in the Prime-Power Case

In this section we will finally prove Karp's conjecture, Conjecture 3.10, in the prime-power case, i.e., we are going to prove that every nontrivial monotone property of graphs with a prime-power number of vertices is evasive.

We give a brief outline of the proof that also serves as a guide through the section. Let $\mathcal{P}$ be a property of graphs on the vertex set V and let $|V|$ be a prime-power. Consider the simplicial complex $K \subseteq \mathcal{P}(X)$ associated with $\mathcal{P}$, where $X = \binom{V}{2}$. We will prove the contrapositive of Karp's conjecture: If $K \neq \emptyset$ and $\mathcal{P}$ is not evasive, then $K = \mathcal{P}(X)$, i.e., $\mathcal{P}$ must be the trivial property containing all graphs. Thinking of K as a simplicial complex, $K = \mathcal{P}(X)$ means that it is the complex given by the simplex X and all its faces.

The nonevasiveness of K yields, by Theorem 3.15, that K is collapsible. Now the symmetry of K inherited by the fact that $\mathcal{P}$ is invariant under graph isomorphisms comes into play. We will consider a symmetry subgroup that acts *transitively* on the vertices of K, i.e., for each pair of vertices of K there is a symmetry interchanging the two vertices.

In the subsequent step, we will employ a strong topological result that states that a contractible simplicial complex with a symmetry group satisfying some group-theoretic condition must contain a simplex that remains fixed under the whole symmetry group. In turn, the transitivity of the group action implies the desired equality $K = \mathcal{P}(X)$.

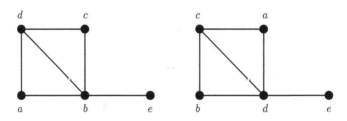

Fig. 3.26 The graphs $G = (V, E)$ and $(V, \pi \cdot E)$

Group Actions on Graph Properties

Let $\mathcal{P}$ be a property of graphs on the vertex set V, i.e., $\mathcal{P} \subseteq \mathcal{P}(X)$, where $X = \binom{V}{2}$ is the set of all edges.

The symmetric group, $\mathrm{Sym}(V)$, acts on the set V via $\pi \cdot u = \pi(u)$, and this action induces an action on the set X of edges via $\pi \cdot uv = \pi(u)\pi(v)$.

Moreover, if $E \subseteq X$ is the edge set of a graph, then $\pi \cdot E$ is defined to be the set of edges

$$\pi \cdot E = \{\pi \cdot uv : uv \in E\} \subseteq X$$

defining a new graph $(V, \pi \cdot E)$. Hence we have an induced action of $\mathrm{Sym}(V)$ on the set of graphs on the vertex set V. Figure 3.26 shows an example of a graph $G = (V, E)$ on the vertex set $V = \{a, b, c, d, e\}$ and its image $(V, \pi \cdot E)$ under the induced action of the element

$$\pi = \begin{pmatrix} a & b & c & d & e \\ b & d & a & c & e \end{pmatrix} \in \mathrm{Sym}(V).$$

The invariance of $\mathcal{P}$ under graph isomorphism translates into the condition that for any $\pi \in \mathrm{Sym}(V)$,

$$E \in \mathcal{P} \text{ if and only if } \pi \cdot E \in \mathcal{P}.$$

In other words, any graph property is invariant under the action of the group $\mathrm{Sym}(V)$.

It is easy to see that this group action is *transitive* on the set X, i.e., if $uv, xy \in \binom{V}{2} = X$ is an arbitrary pair of edges, then there exists a $\pi \in \mathrm{Sym}(V)$ such that $\pi \cdot uv = xy$.

Lemma 3.16. *Let $K \subseteq \mathcal{P}(X)$ be a simplicial complex. Assume that the group G acts transitively on X such that K is invariant under the induced action. If the set of fixed points $|K|^G$ of the induced action on the polyhedron $|K|$ is nonempty, then $K = \mathcal{P}(X)$.*

Proof. Let $x \in |K|^G$ be a fixed point and $\sigma \in K$ the simplex that is minimal under inclusion with the property that $x \in |\sigma|$. We claim that $\sigma = X$, which proves the lemma. Since σ cannot be a vertex by the transitivity of the group action, x is, in fact, contained in the interior of $|\sigma|$. If $\sigma \neq X$, then choose elements $a \in \sigma$ and $b \in X \setminus \sigma$, and let $g \in G$ be such that $g \cdot a = b$. Then clearly $g \cdot \sigma \neq \sigma$, and hence in particular, $\mathrm{int}(g \cdot |\sigma|) \cap \mathrm{int}(|\sigma|) = \mathrm{int}(|g \cdot \sigma|) \cap \mathrm{int}(|\sigma|) = \emptyset$. This contradicts the fact that $|\sigma|$ contains the fixed point x.   □

The following theorem from Smith theory—the homological theory of orbits and fixed points of group actions on simplicial complexes—is an immediate corollary of Theorem E.16 on page 228.

Theorem 3.17. *Let $K \subseteq \mathcal{P}(X)$ be a simplicial complex. Assume that the finite group G acts on X such that K is invariant under the induced action. Furthermore, assume that*

- *$|K|$ is contractible,*
- *There exists a normal subgroup $H \trianglelefteq G$, such that H is a p-group, i.e., the order of H is a power of a prime p, and*
- *The quotient G/H is cyclic.*

Then the set of fixed points $|K|^G$ is nonempty.   □

A Group Action in the Prime-Power Case

Assume now that the number of elements of the vertex set V is a prime-power. Let's say $|V| = p^r$ for some prime p. Without loss of generality, we may assume that $V = F$ is the ground set of the finite Galois field F with p^r elements [Hun74]. Consider the following subgroup G of $\mathrm{Sym}(V)$:

$$G = \{f_{a,b} : a, b \in F, a \neq 0\},$$

where $f_{a,b} : V \to V$ is the affine linear function defined by $f_{a,b}(x) = ax + b$. Then G acts transitively on $\binom{V}{2}$ due to the fact that $\det \begin{pmatrix} u & 1 \\ v & 1 \end{pmatrix} = u - v$ is nonzero for $u \neq v$. Now consider the subgroup

$$H = \{f_{0,b} : b \in F\}$$

of G. Clearly H is a p-group, since its cardinality $|H| = |F| = p^r$ is a power of a prime. Moreover, the quotient G/H is isomorphic to the multiplicative group $(F \setminus \{0\}, \cdot)$ of the field F. It is a basic exercise in algebra to see that this group is always cyclic for finite fields F.

A Proof of Karp's Conjecture in the Prime-Power Case

Now we have only to put the pieces together in order to obtain a proof of Karp's conjecture in the prime-power case.

Theorem 3.18 (Khan, Saks, Sturtevant [KSS84]). *Assume that $\mathcal{P}$ is a nontrivial monotone graph property of graphs on the vertex set V and that $|V| = p^r$ is a power of a prime. Then $\mathcal{P}$ is evasive.*

Proof. Assume that $\mathcal{P} \neq \emptyset$ is a monotone nonevasive property of graphs on the vertex set V. We have to show that $\mathcal{P} = \mathcal{P}(X)$ is the remaining trivial property. Let $X = \binom{V}{2}$ and let G be defined as above, acting transitively on X. By Theorem 3.15, $\mathcal{P}$ is collapsible. Since collapsible complexes are contractible, the desired result follows from Theorem 3.17 and Lemma 3.16.   □

3.4   The Rivest–Vuillemin Theorem on Set Systems

We end this chapter with a theorem by Rivest and Vuillemin [RV76] that is very similar to Theorem 3.18 and proves evasiveness for general set systems on a ground set of prime-power cardinality with a transitive group action. In contrast to Theorem 3.18, it has an elementary proof based purely on a counting argument.

Let X be a set and $S \subseteq \mathcal{P}(X)$ a family of subsets. The starting point is a very interesting observation, the proof of which we leave as an exercise.

Lemma 3.19. *If the number of sets in S of even cardinality is different from the number of sets of odd cardinality, then (S, X) is evasive.*   □

Now assume that a group $G \leq \text{Sym}(X)$ acts transitively on X and leaves S invariant, i.e., $\pi \cdot S = \{\pi(\sigma) : \sigma \in S\} = S$ for every $\pi \in G$. Let $\sigma \in S$ be an element of S and $G \cdot \sigma = \{\sigma_1, \dots, \sigma_k\} \subseteq S$ the orbit of σ. For $x \in X$, let $h(x)$ be defined to be the number of sets in the orbit that contain x, i.e.,

$$h(x) = |\{i \in [k] : x \in \sigma_i\}|.$$

By the transitivity of the group action, we easily see that $h(x)$ is independent of the choice of $x \in X$. So let us denote this number simply by h. Then, by double counting, we obtain the identity $k|\sigma| = h|X|$.

Theorem 3.20. *Let X be a set of prime-power cardinality p^r and $S \subseteq \mathcal{P}(X)$ a family of subsets such that $\emptyset \in S$ and $X \notin S$. If, moreover, there exists a transitive group action on X leaving S invariant, then (S, X) is evasive.*

Proof. Let $\sigma \in S$, $\sigma \neq \emptyset$, and let $G \cdot \sigma = \{\sigma_1, \dots, \sigma_k\} \subseteq S$ be the corresponding orbit. Then, by the preceding considerations, $k|\sigma| = hp^r$. Since $|\sigma| < p^r$, we conclude that p divides the size k of the orbit. Hence p divides the size of each orbit in S except the size of the orbit $\{\emptyset\}$. Since S is partitioned by the orbits, the number of sets in S of even cardinality turns out to be different from the number of sets of odd cardinality. $\square$

Note that this beautiful theorem has no effect on Karp's conjecture since the number of edges of a complete graph K_n is a prime-power only in the case $n = 3$.

Exercises

1. Let (a_{ij}) be a matrix with real entries. Show that

$$\max_i \min_j a_{ij} \le \min_j \max_i a_{ij}$$

   with equality if and only if there exist i_0 and j_0 such that the entry $a_{i_0 j_0}$ is minimal in row i_0 and maximal in column j_0.

2. Give an example of a graph property $\mathcal{P}$ satisfying

$$\min_\phi \max_G c(\mathcal{P}, \phi, \psi_G) \ne \max_G \min_\phi c(\mathcal{P}, \phi, \psi_G).$$

3. Let $I = [0, 1]$ be the unit interval. Prove or disprove: if $f : I \times I \longrightarrow \mathbb{R}$ is a continuous map, then

$$\max_{x \in I} \min_{y \in I} f(x, y) = \min_{y \in I} \max_{x \in I} f(x, y).$$

4. Prove the converse of Lemma 3.5. More precisely, show that if ψ is the greedy strategy for the hider and if

$$\min_\phi c(\mathcal{P}, \phi, \psi) = \binom{n}{2},$$

   then for each $E \in \mathcal{P}$ and $e \in E$ with $E \setminus \{e\} \in \mathcal{P}$, there exist an $f \in \binom{V}{2} \setminus E$ and $F \in \mathcal{P}$ such that $(E \setminus \{e\}) \cup \{f\} \subseteq F$.

5. Give a proof of Theorem 3.6 on page 75.

6. Provide the missing details for the case $|I| = 2$ in Phase I of the algorithm in the proof of Theorem 3.8, thereby completing the proof.

7. Prove that $c(\mathcal{P}) = c(\overline{\mathcal{P}})$ for any graph property $\mathcal{P}$, where $\overline{\mathcal{P}}$ is as defined on page 82.

8. Show that the graph property $\mathcal{B}_6$ defined on page 77 is not evasive.

9. Property $\mathcal{B}_6$ can easily be generalized to a property of graphs on a fixed number $n \ge 6$ of vertices. Let $\mathcal{B}_n$ be the property given by all graphs on n vertices isomorphic to any of the three shown in Fig. 3.27. Compared to Fig. 3.10, the center edge has been replaced by a path of length $n - 5$.
   Show that the graph property $\mathcal{B}_n$ is not evasive.

Fig. 3.27 The three isomorphism types of graphs in $\mathcal{B}_n$

Fig. 3.28 Graphs describing a nonevasive graph property

10. Let $\mathcal{P}$ be the graph property given by all possible graphs that are isomorphic to one of the three graphs shown in Fig. 3.28. Show that $\mathcal{P}$ does not satisfy the condition of Lemma 3.5.

11. Show that the graph property $\mathcal{P}$ defined in the previous exercise is not evasive [MW76].

12. Show that if K' is obtained from K by an elementary collapse, i.e., $K \searrow K'$, then the polyhedron $|K'|$ is a strong deformation retract of $|K|$. In particular, the polyhedra $|K|$ and $|K'|$ are homotopy equivalent.

13. Provide the details of the proof of Lemma 3.14 on page 87.

14. Show that the group $G = \{f_{a,b} : a, b \in F, a \neq 0\} \leq \mathrm{Sym}(F)$ as defined on page 91 acts transitively on $\binom{F}{2}$ for any finite field F.

15. Let G be defined as in the previous exercise and $H = \{f_{0,b} : b \in F\} \leq G$. Show that the quotient G/H is isomorphic to the multiplicative group $(F \setminus \{0\}, \cdot)$.

16. Let $(G, \cdot, e)$ be a finite multiplicative group and m the largest order among its elements. Show that $g^m = e$ for any $g \in G$.

17. Let F be a finite field. Use the previous exercise to show that the multiplicative group $(F \setminus \{0\}, \cdot)$ is cyclic. Hint: Consider the polynomial $x^m - 1 \in F[x]$.

18. This exercise is concerned with a result by Kahn, Saks, and Sturtevant [KSS84] showing asymptotic quadratic complexity for nontrivial monotone graph properties. Let

$$c(n) = \min\{c(\mathcal{P}) : \mathcal{P} \text{ monotone, nontrivial property of } n\text{-vertex graphs}\}.$$

Prove that $c(n) \geq \frac{n^2}{4} - \varphi(n)$, where $\varphi : \mathbb{N} \to \mathbb{N}$ is a function with $\lim_{n \to \infty} \frac{\varphi(n)}{n^2} = 0$.

Besides Theorem 3.18, you may use the following results.

(a) A lemma by Kleitman and Kwiatkowski [KK80] stating that

$$c(n) \geq \min\{c(n-1), q(n-q)\},$$

where q is the prime-power nearest to $\frac{n}{2}$.

(b) A result from number theory that states that there is a function $\psi : \mathbb{N} \to \mathbb{N}$ with $\lim_{n\to\infty} \frac{\psi(n)}{n} = 0$ and the property that for each n there exists a prime number between $n - \psi(n)$ and $n + \psi(n)$.

19. Provide a proof of Lemma 3.19 on page 92.

20. Show that $h(x)$, as defined on page 92, is independent of the choice of $x \in X$.

Chapter 4
Embedding and Mapping Problems

Embedding problems in discrete geometry lead to very challenging and interesting questions. A very basic question is whether a given graph G is planar, i.e., can be drawn in the plane such that edges do not cross. Kuratowski's theorem, Theorem A.18, answers this question in terms of forbidden subgraphs. In this chapter we want to pursue an alternative characterization with methods from algebraic topology that reduce to simple linear algebra computations.

More generally, we may ask whether a given simplicial complex admits a geometric realization in some fixed dimension n. As shown in Proposition B.41 on page 177, any d-dimensional simplicial complex admits a geometric realization in $\mathbb{R}^{2d+1}$. Is the dimension $2d + 1$ optimal in general? We will introduce a general criterion in order to investigate these questions.

Beyond questions of embeddability and nonembeddability, we will be concerned with whether, for a given graph or complex, all maps into some Euclidean space have some predetermined intersection property and whether maps with such a property exist.

4.1 The Radon Theorems

We start with a classical theorem from convex geometry: the affine Radon theorem.

Theorem 4.1. *Any set $S \subseteq \mathbb{R}^d$ of $d + 2$ points admits a partition $S = U \dot\cup V$ into two sets such that their convex hulls intersect nontrivially, i.e., $\mathrm{conv}(U) \cap \mathrm{conv}(V) \neq \emptyset$.*

Figure 4.1 shows the essentially different configurations of points, and the resulting partitions, in the case $d = 2$.

The affine Radon theorem is, in fact, a very special nonembeddability result. This becomes clear in the following equivalent version about affine linear maps of the standard simplex into Euclidean space.

M. de Longueville, *A Course in Topological Combinatorics*, Universitext, DOI 10.1007/978-1-4419-7910-0_4, © Springer Science+Business Media New York 2013

Fig. 4.1 Three configurations of four points in $\mathbb{R}^2$ and intersecting convex hulls

Theorem 4.2. *For any affine linear map* $f : \sigma^{d+1} \longrightarrow \mathbb{R}^d$, *there exist two disjoint faces* $\tau_1, \tau_2 \leq \sigma^{d+1}$ *such that* $f(\tau_1) \cap f(\tau_2) \neq \emptyset$.

Observe that since τ_1 and τ_2 must be nonempty, this theorem may also be stated for affine linear maps $\partial\sigma^{d+1} \to \mathbb{R}^d$ from the boundary of the simplex to Euclidean space. In this form, it may remind the reader of a version of the Borsuk–Ulam theorem. This connection is investigated in [Bár79]. In particular, one is tempted to wonder whether affine linearity is necessary or whether the condition on the map can be weakened. This leads us to the continuous Radon theorem.

Theorem 4.3. *For any continuous map* $f : \sigma^{d+1} \longrightarrow \mathbb{R}^d$, *there exist two disjoint faces* $\tau_1, \tau_2 \leq \sigma^{d+1}$ *such that* $f(\tau_1) \cap f(\tau_2) \neq \emptyset$.

Indeed, this theorem can be proven by applying the Borsuk–Ulam theorem, e.g., by supplying a continuous map $g : \mathbb{S}^d \to \sigma^{d+1}$ such that for every $x \in \mathbb{S}^d$, the minimal faces of σ^{d+1} containing $g(x)$ and $g(-x)$ are disjoint.

We will use the continuous Radon theorem as a motivating example for developing a method to obtain nonembeddability results. The continuous—and hence also the affine linear—Radon theorem will then be proven as a corollary to Proposition 4.10 on page 101.

4.2 Deleted Joins and the $\mathbb{Z}_2$-Index

Deleted Join of Spaces

In order to attempt a methodical approach, we want to start with a fairly general embedding problem. Let X and Y be topological spaces. In our situation, X will be the polyhedron of a simplicial complex—as defined on page 175—and $Y = \mathbb{R}^d$. We want to show that there is no embedding from X into Y, i.e., there does not exist a map $f : X \to Y$ that is a homeomorphism onto its image. In order to prove this, it suffices to show that there is no continuous injective map $f : X \to Y$. Assuming that a continuous injective map f exists, it induces a map of the twofold joins

$$f^{*2} : X * X \longrightarrow Y * Y,$$

$$(tx_1, (1-t)x_2) \longmapsto (tf(x_1), (1-t)f(x_2)).$$

Injectivity of f translates into the condition that

$$f^{*2}\left(\frac{1}{2}x_1, \frac{1}{2}x_2\right) \notin \left\{\left(\frac{1}{2}y, \frac{1}{2}y\right) : y \in Y\right\}$$

for every $x_1 \neq x_2$. In other words, we obtain a restricted map

$$f_\Delta^{*2} : X * X \setminus \left\{\left(\frac{1}{2}x, \frac{1}{2}x\right) : x \in X\right\} \longrightarrow Y * Y \setminus \left\{\left(\frac{1}{2}y, \frac{1}{2}y\right) : y \in Y\right\}.$$

We call the space $X * X \setminus \left\{\left(\frac{1}{2}x, \frac{1}{2}x\right) : x \in X\right\}$ the (twofold) *deleted join of* X, which we denote by $X *_\Delta X$. Observe that $\mathbb{Z}_2$ acts freely on the deleted join via interchanging the two coordinates

$$(tx_1, (1 - t)x_2) \mapsto ((1 - t)x_2, tx_1),$$

and clearly the map f_Δ^{*2} is equivariant with respect to this action.

Hence, in order to obtain nonembeddability of X into Y, it suffices to show that there is no $\mathbb{Z}_2$-equivariant map $X *_\Delta X \to Y *_\Delta Y$.

$\mathbb{Z}_2$-*Index*

An appropriate tool in order to decide the nonexistence of $\mathbb{Z}_2$-equivariant maps is the $\mathbb{Z}_2$-index. Recall Definition 2.19 from page 59. If X is a topological space with a $\mathbb{Z}_2$-action, then the $\mathbb{Z}_2$-*index of* X is defined to be

$$\mathrm{ind}(X) = \min \left\{k \geq 0 : \text{there exists a } \mathbb{Z}_2\text{-map } X \to \mathbb{S}^k\right\}.$$

An immediate observation is that if $f : X \to Y$ is an equivariant map of $\mathbb{Z}_2$-spaces and $g : Y \to \mathbb{S}^k$ witnesses $\mathrm{ind}(Y) = k$, then the composition $g \circ f : X \to \mathbb{S}^k$ is $\mathbb{Z}_2$-equivariant and yields $\mathrm{ind}(X) \leq k = \mathrm{ind}(Y)$. In other words, we have observed the following fact, which provides us with a tool that fits our needs.

Lemma 4.4. *If* $\mathrm{ind}(X) > \mathrm{ind}(Y)$, *then there is no* $\mathbb{Z}_2$-*equivariant map* $X \to Y$.

Since we are addressing the question of embeddings into $\mathbb{R}^d$, we should determine the index of the deleted join $\mathbb{R}^d *_\Delta \mathbb{R}^d$. For our purposes it suffices to bound the index from above.

Proposition 4.5. *The* $\mathbb{Z}_2$-*index of* $\mathbb{R}^d *_\Delta \mathbb{R}^d$ *satisfies the bound* $\mathrm{ind}(\mathbb{R}^d *_\Delta \mathbb{R}^d) \leq d$.

Proof. For $x = (x_1, \ldots, x_d) \in \mathbb{R}^d$, let $\bar{x} = (1, x_1, x_2, \ldots, x_d) \in \mathbb{R}^{d+1}$. Then the map

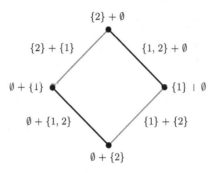

Fig. 4.2 Deleted join of a 1-simplex

$$\mathbb{R}^d *_\Delta \mathbb{R}^d \longrightarrow \mathbb{S}^d,$$

$$(tx, (1-t)y) \longmapsto \frac{t\bar{x} - (1-t)\bar{y}}{\|t\bar{x} - (1-t)\bar{y}\|},$$

is a continuous $\mathbb{Z}_2$-equivariant map.   □

In fact, with a little more effort it can be shown that $\mathrm{ind}(\mathbb{R}^d *_\Delta \mathbb{R}^d) = d$.

Deleted Join of Simplicial Complexes

In the sequel we will need a combinatorial analogue of the deleted join operation for simplicial complexes.

Definition 4.6. The *(twofold) deleted join* $K_\Delta^{*2} = K *_\Delta K$ *of a simplicial complex* K *is defined as the complex*

$$K_\Delta^{*2} = \{\tau_1 + \tau_2 : \tau_1, \tau_2 \in K, \tau_1 \cap \tau_2 = \emptyset\} \subseteq K * K.$$

Let's develop some intuition for the deleted join. First of all, note that $\mathbb{Z}_2$ acts freely on the deleted join via $\tau_1 + \tau_2 \mapsto \tau_2 + \tau_1$. The deleted join of a simplex is easily determined, and its geometric realization turns out to be a sphere. An example is given in Fig. 4.2.

Lemma 4.7. *Let* K *be the abstract simplicial complex representing an n-simplex and its faces, i.e.,* $K = 2^{[n+1]} = \{\tau : \tau \subseteq [n+1]\}$. *Then the deleted join* $K *_\Delta K$ *is* $\mathbb{Z}_2$-*equivariant isomorphic to the boundary complex* $K(\Gamma^n)$ *of the* $(n+1)$-*dimensional cross polytope (as defined on page 14). In particular,* $\mathrm{ind}(|K *_\Delta K|) = n$.

Proof. The following map clearly yields an isomorphism of complexes:

$$K *_\Delta K \longrightarrow K(\Gamma^n),$$

$$\tau_1 + \tau_2 \longmapsto \{e_i : i \in \tau_1\} \cup \{-e_i : i \in \tau_2\}. \qquad \square$$

The previous lemma can also be obtained via the following observation. The details are left to the exercises.

Lemma 4.8. *If K and L are simplicial complexes, then there is an isomorphism* $(K * L)_\Delta^{*2} \cong K_\Delta^{*2} * L_\Delta^{*2}$.

Proof. The isomorphism is given by

$$(K * L)_\Delta^{*2} \longmapsto K_\Delta^{*2} * L_\Delta^{*2},$$

$$(\sigma_1 + \tau_1) + (\sigma_2 + \tau_2) \longmapsto (\sigma_1 + \sigma_2) + (\tau_1 + \tau_2). \qquad \square$$

Maps and Deleted Joins

Now let us return to the continuous Radon theorem, now phrased in a slightly more general setting. Let $d \geq 1$, and let K be an (abstract) simplicial complex for which we want to show that for any continuous map $f : |K| \to \mathbb{R}^d$, there exist disjoint faces $\tau_1, \tau_2 \in K$ such that $f(|\tau_1|) \cap f(|\tau_2|) \neq \emptyset$.

Note 4.9. Any continuous map $f : |K| \to \mathbb{R}^d$ induces the following composition of continuous $\mathbb{Z}_2$-equivariant maps:

$$|K *_\Delta K| \hookrightarrow |K * K| \overset{\cong}{\longrightarrow} |K| * |K| \overset{f^{*2}}{\longrightarrow} \mathbb{R}^d * \mathbb{R}^d.$$

If we now assume to the contrary that $f : |K| \to \mathbb{R}^d$ is a map with $f(x_1) \neq f(x_2)$ whenever $x_1 \in |\tau_1|, x_2 \in |\tau_2|$ and $\tau_1 \cap \tau_2 = \emptyset$, then f induces a $\mathbb{Z}_2$-equivariant map

$$|K *_\Delta K| \longrightarrow \mathbb{R}^d *_\Delta \mathbb{R}^d.$$

Hence we obtain the inequality $\mathrm{ind}(|K *_\Delta K|) \leq \mathrm{ind}(\mathbb{R}^d *_\Delta \mathbb{R}^d) \leq d$ and obtain the following result.

Proposition 4.10. *Let K be a simplicial complex. If $\mathrm{ind}(|K *_\Delta K|) > d$, then for every continuous map $f : |K| \to \mathbb{R}^d$, there exist disjoint faces $\tau_1, \tau_2 \in K$ such that $f(|\tau_1|) \cap f(|\tau_2|) \neq \emptyset$.* $\qquad \square$

Corollary 4.11 (Continuous Radon theorem). *For any continuous map $f : \sigma^{d+1} \longrightarrow \mathbb{R}^d$, there exist two disjoint faces $\tau_1, \tau_2 \leq \sigma^{d+1}$ such that $f(\tau_1) \cap f(\tau_2) \neq \emptyset$.*

Proof. Let $K = \{\tau : \tau \subseteq [d+2]\}$ be the abstract simplicial complex with geometric realization σ^{d+1}. Then Lemma 4.7 implies $\mathrm{ind}(|K *_\Delta K|) = d + 1 > d$. $\qquad \square$

4.3   Bier Spheres

We want to apply Proposition 4.10 in a more sophisticated situation. In order to do so, we will investigate a certain family of spheres.

With any proper simplicial subcomplex K of the abstract $(n-1)$ simplex $2^{[n]}$, one can associate a certain triangulated $(n-2)$-dimensional sphere, $\mathrm{Bier}_n(K)$, known as the *Bier sphere* [Bie92].

In this section we will explain the construction of the Bier spheres and will prove that these complexes are indeed spheres. It is based on the article [Lon04].

The Combinatorial Alexander Dual and Barycentric Subdivision

Let $K \subset 2^{[n]}$ be a proper subcomplex of the simplex. The *combinatorial Alexander dual $K^\star$ of K* is given by the complements of the nonfaces of K:

$$K^\star = \{\tau \subset [n] : \overline{\tau} \notin K\},$$

where $\overline{\tau}$ denotes the complement $[n] \setminus \tau$. One way to characterize the Alexander dual is by the property that its facets are given by the complements of the minimal nonfaces of K.

Recall that for a simplicial complex L, the simplices of its first barycentric subdivision, $\mathrm{sd}(L)$, are given by (possibly empty) inclusion chains of nonempty simplices of L. The following immediate lemma characterizes the Alexander dual via its first barycentric subdivision.

Lemma 4.12.   *For any proper subcomplex $K \subset 2^{[n]}$, we have*

$$\{t_0 \subset \cdots \subset t_l\} \in \mathrm{sd}(K^\star) \subseteq \mathrm{sd}(2^{[n]} \setminus \{[n]\})$$

if and only if $\{\overline{t_l} \subset \cdots \subset \overline{t_0}\}$ is disjoint from $\mathrm{sd}(K) \subseteq \mathrm{sd}\left(2^{[n]} \setminus \{[n]\}\right)$.   □

An illustration of the lemma is given in Fig. 4.3. In this case, the minimal nonfaces of K are given by the simplices $\{1, 4\}$ and $\{2, 3, 4\}$. Now consider, for example, the simplex $\{t_0 \subset t_1\} = \{\{3\} \subset \{2, 3\}\} \in \mathrm{sd}(K^\star)$. Then $\{\overline{t_1} \subset \overline{t_0}\} = \{\{1, 4\} \subset \{1, 2, 4\}\}$ is disjoint from $\mathrm{sd}(K)$.

Shore Subdivisions

We are interested in a certain type of subdivision of subcomplexes of a join, $K * L$, of complexes. The following observation is immediate.

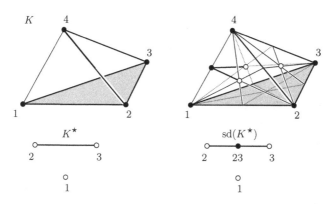

Fig. 4.3 A complex K, its Alexander dual K^*, and the interplay of their barycentric subdivisions

Lemma 4.13. *Let $K \subseteq 2^{[n]}$ and $L \subseteq 2^{[m]}$ be two abstract simplicial complexes, and let J be a subcomplex of the join $K * L$. Then J and the complex*

$$\{s * t : s \in \mathrm{sd}(2^\sigma), t \in \mathrm{sd}(2^\tau), \sigma * \tau \in J\} \subseteq \mathrm{sd}(K) * \mathrm{sd}(L)$$

possess identical geometric realizations. □

In the literature, the complex $\{s * t : s \in \mathrm{sd}(2^\sigma), t \in \mathrm{sd}(2^\tau), \sigma * \tau \in J\}$ is also referred to as the *shore subdivision of J with respect to $K * L$*. Note that in general, it is distinct from $\mathrm{sd}(J)$.

The Bier Sphere

For any proper subcomplex $K \subset 2^{[n]}$, the *Bier sphere*, $\mathrm{Bier}_n(K)$, is defined as the deleted join of K with its combinatorial Alexander dual, i.e.,

$$\mathrm{Bier}_n(K) = K *_\Delta K^* = \{\sigma * \tau : \sigma \in K, \tau \in K^*, \sigma \cap \tau = \emptyset\}$$

$$= \{\sigma * \tau : \sigma \in K, \overline{\tau} \notin K, \sigma \subseteq \overline{\tau}\}.$$

Lemma 4.12 and Fig. 4.3 suggest the following.

Proposition 4.14. *For any proper subcomplex $K \subset 2^{[n]}$, the geometric realization of the Bier sphere, $|\mathrm{Bier}_n(K)|$, is an $(n-2)$-sphere. More precisely, the shore subdivision of $\mathrm{Bier}_n(K)$ with respect to $K * K^*$ is isomorphic to $\mathrm{sd}(2^{[n]} \setminus \{[n]\})$, i.e., the barycentric subdivision of the boundary of the $(n-1)$-dimensional simplex.*

Proof. The shore subdivision of $\mathrm{Bier}_n(K)$ is given by

$$\{s * t : s \in \mathrm{sd}(2^\sigma), t \in \mathrm{sd}(2^\tau), \sigma * \tau \in K *_\Delta K^\star\}$$
$$= \{(s_1 \subset \cdots \subset s_k) * (t_1 \subset \cdots \subset t_l) : \emptyset \neq s_i \subseteq \sigma, \emptyset \neq t_j \subseteq \tau, \sigma \in K, \overline{\tau} \notin K, \sigma \subseteq \overline{\tau}\}.$$

It maps isomorphically to $\mathrm{sd}(2^{[n]} \setminus \{[n]\})$ via

$$(s_1 \subset \cdots \subset s_k) * (t_1 \subset \cdots \subset t_l) \mapsto (s_1 \subset \cdots \subset s_k \subset \overline{t_l} \subset \cdots \subset \overline{t_1}).$$

This map is well defined and injective, since $s_k \subseteq \sigma \subseteq \overline{\tau} \subseteq \overline{t_l} \subset [n]$. To see surjectivity, let $u_1 \subset \cdots \subset u_m \in \mathrm{sd}(2^{[n]} \setminus \{[n]\})$. Let $u_0 = \emptyset$, $u_{m+1} = [n]$, and $k \geq 0$ be maximal with the property that $u_k \in K$. Then define σ to be u_k and τ to be $\overline{u_{k+1}}$. Set $s_1 = u_1, \ldots, s_k = u_k$ and $t_1 = \overline{u_m}, \ldots, t_{m-k} = \overline{u_{k+1}}$. Then trivially the following hold:

- $s_1 \subset \cdots \subset s_k \subseteq \sigma$,
- $t_1 \subset \cdots \subset t_{m-k} \subseteq \tau$,
- $\sigma \in K, \overline{\tau} \notin K, \sigma \subset \overline{\tau}$, and
- $(s_1 \subset \cdots \subset s_k) * (t_1 \subset \cdots \subset t_{m-k}) \mapsto u_1 \subset \cdots \subset u_m$.

An application of Lemma 4.13 finishes the proof. $\square$

4.4   The van Kampen–Flores Theorem

Now we return to the question of realizability of simplicial complexes. Proposition B.41 shows that any d-dimensional simplicial complex can geometrically be realized in $\mathbb{R}^{2d+1}$. The following van Kampen–Flores theorem shows that in general, this is optimal.

Theorem 4.15. *Let K be the d-dimensional skeleton of the abstract $(2d + 2)$-dimensional simplex, i.e.,*

$$K = \{\sigma : \sigma \subseteq [2d + 3], |\sigma| \leq d + 1\}.$$

Then there is no embedding from $|K|$ into $\mathbb{R}^{2d}$. In fact, for any continuous map $f : |K| \to \mathbb{R}^{2d}$, there exist two disjoint simplices of K whose images intersect.

Note that the special case $d = 1$ says that the 1-dimensional skeleton of the 4-dimensional simplex is not embeddable in $\mathbb{R}^2$, i.e., the complete graph K_5 on five vertices is not planar.

Proof. We want to combine the concept of Bier spheres with Proposition 4.10. First we compute the combinatorial Alexander dual of K and obtain a surprising result:

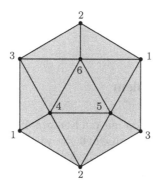

Fig. 4.4 A six-vertex triangulation of the projective plane

$$K^\star = \{\tau \subseteq [2d+3] : [2d+3] \setminus \tau \notin K\}$$
$$= \{\tau \subseteq [2d+3] : |[2d+3] \setminus \tau| \geq d+2\}$$
$$= \{\tau \subseteq [2d+3] : |\tau| \leq d+1\} = K.$$

Hence for the twofold deleted join, we obtain

$$K *_\Delta K = K *_\Delta K^\star = \mathrm{Bier}_{2d+3}(K),$$

which is a $(2d+1)$-dimensional sphere with $\mathbb{Z}_2$-index $2d+1$. $\square$

A Six-Vertex Triangulation of the Projective Plane

There is another interesting example of a simplicial complex K with the property that it coincides with its Alexander dual $K^\star$. Consider the abstract simplicial complex K on the vertex set $[6]$ as given in Fig. 4.4.

The polyhedron of its geometric realization is a projective plane. Its facets are given by the triples

$$124, 126, 134, 135, 156, 235, 236, 245, 346, 456.$$

Note that K has $15 = \binom{6}{2}$ edges, i.e., the complete graph on six vertices is embedded in $|K|$. Hence the minimal nonfaces are 2-dimensional. It is now easy to check that $\sigma \in K$ if and only if $[6] \setminus \sigma \notin K$ for any $\sigma \in \binom{[6]}{3}$. And hence K coincides with its Alexander dual $K^\star$.

We obtain that $K *_\Delta K = K *_\Delta K^\star = \mathrm{Bier}_6(K)$ is a 4-dimensional sphere, and hence by Proposition 4.10, the polyhedron $|K|$ may not be embedded into $\mathbb{R}^3$.

Proposition 4.16. *There is no embedding of the projective plane into 3-dimensional Euclidean space.*   □

4.5 The Tverberg Problem

The previously considered Radon and Van Kampen–Flores theorems are essentially nonembeddability results. But because of the way they are phrased, they are actually stronger, since they predict disjoint simplices whose images intersect. In this section we want to discuss a very strong generalization of the Radon theorem.

As in the evolution of the Radon theorems, the Tverberg conjecture has its origin in an affine version that reduces to the affine Radon theorem in the case $q = 2$.

Theorem 4.17 (Affine Tverberg). *Let $d \geq 1$ and $q \geq 2$ and set $N = (d + 1)(q-1)$. For any set $S \subseteq \mathbb{R}^d$ of $N+1$ points, there exists a partition $S = S_1 \dot{\cup} \cdots \dot{\cup} S_q$ such that the convex hulls of the S_i intersect nontrivially, i.e., $\mathrm{conv}(S_1) \cap \cdots \cap \mathrm{conv}(S_q) \neq \emptyset$.*   □

The statement of the theorem was conjectured by Bryan J. Birch [Bir59] in 1959 and was established by Helge Tverberg [Tve66] in 1966. A very beautiful proof employing elementary algebraic methods was given by Karanbir S. Sarkaria [Sar92] in 1992.

To get a feeling for this result, we turn our attention to an open conjecture by Gerard Sierksma. He conjectured that the number of different *Tverberg partitions* $S = S_1 \dot{\cup} \cdots \dot{\cup} S_q$ is at least $((q - 1)!)^d$. This number is attained for the following *Sierksma configuration* of $N + 1 = (q - 1)(d + 1) + 1$ points. Consider the $d + 1$ vertices of a simplex in $\mathbb{R}^d$. For each vertex of the simplex, place $q - 1$ points in an epsilon neighborhood of the vertex. The final point is placed at the barycenter of the simplex. A Tverberg partition is then obtained as follows. Totally order each set of points in the epsilon neighborhoods, and for each $1 \leq i \leq q - 1$, let S_i consist of the $d + 1$ points that occur in position i of these orderings. Define S_q to be the point at the barycenter. For an illustration we refer to Fig. 4.5. It is not hard to show that all Tverberg partitions are obtained in this way and that there are exactly $((q - 1)!)^d$ of them.

As with the affine Radon theorem, the affine Tverberg theorem may be rephrased as a statement about affine linear maps $f : \sigma^N \to \mathbb{R}^d$. Such a reformulation asks for the following natural generalization.

Conjecture 4.18 (Continuous Tverberg). Let $d \geq 1$ and $q \geq 2$ and set $N = (d + 1)(q - 1)$. For any continuous map $f : \sigma^N \to \mathbb{R}^d$, there exist pairwise disjoint faces $\tau_1, \ldots, \tau_q \leq \sigma^N$ such that their images under f intersect nontrivially, i.e., $f(\tau_1) \cap \cdots \cap f(\tau_q) \neq \emptyset$.

This conjecture was proved in the case that q is a prime number by Imre Bárány, Senya B. Schlosman, and András Szücs in 1981 [BSS81]. The next step, the case that q is a prime power, was taken by Murad Özaydin in 1987, Alexey Yu. Volovikov in 1996, and Karanbir S. Sarkaria in 2000.

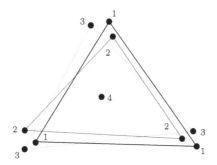

Fig. 4.5 The Sierksma configuration for $d = 2$ and $q = 4$

The general conjecture is still open. Here we are going to prove the prime-power case by applying Theorem 1.17.

Theorem 4.19 ([Öza87, Vol96, Sar00]). *Let $d \geq 1$ and let $q = p^r$ be a prime power, and set $N = (d + 1)(q - 1)$. For any continuous map $f : \sigma^N \to \mathbb{R}^d$, there exist pairwise disjoint faces $\tau_1, \ldots, \tau_q \leq \sigma^N$ such that their images under f intersect nontrivially, i.e., $f(\tau_1) \cap \cdots \cap f(\tau_q) \neq \emptyset$.*

Before we start with the proof of this theorem, we need to introduce the concept of q-fold deleted joins in order to have a useful concept to deal with the sets $\tau_1, \ldots, \tau_q \leq \sigma^N$ of pairwise disjoint faces.

q-Fold Deleted Joins

Let K be an abstract simplicial complex. Generalizing the construction of the twofold deleted join K_Δ^{*2}, we may define for any natural number q the *q-fold deleted join* K_Δ^{*q} to be the following subcomplex of the q-fold join:

$$K_\Delta^{*q} = \{\tau_1 + \cdots + \tau_q : \tau_1, \ldots, \tau_q \in K, \forall i, j \in [q] : \tau_i \cap \tau_j = \emptyset\}.$$

We are mostly interested in the case that $K = 2^{[N+1]} = \{\tau : \tau \subseteq [N + 1]\}$ is an N-simplex. To this end, let Θ_q be the simplicial complex with q vertices and no higher-dimensional simplices, i.e.,

$$\Theta_q = \{\emptyset, \{s_1\}, \ldots, \{s_q\}\}.$$

Lemma 4.20. *Let $q \geq 1, N \geq 0$. Then there is an isomorphism of simplicial complexes $(2^{[N+1]})_\Delta^{*q} \cong (\Theta_q)^{*(N+1)}$.*

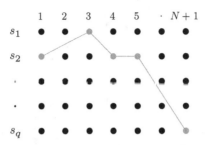

Fig. 4.6 An illustration of $(2^{[N+1]})_\Delta^{*q} \cong (\Theta_q)^{*(N+1)}$

Proof. The isomorphism is given by the map

$$(2^{[N+1]})_\Delta^{*q} \longrightarrow (\Theta_q)^{*(N+1)},$$

$$\tau_1 + \cdots + \tau_q \longmapsto \sigma_1 + \cdots + \sigma_{N+1},$$

where for all $i = 1, \ldots, N + 1$,

$$\sigma_i = \begin{cases} \{s_j\}, & \text{if } i \in \tau_j, \\ \emptyset, & \text{otherwise.} \end{cases}$$

This is illustrated in Fig. 4.6, where the simplices τ_j and σ_i are easily read from rows and columns, respectively. In the figure we have $\tau_1 = \{3\}$, $\tau_2 = \{1, 4, 5\}$, $\tau_q = \{N + 1\}$, and all other τ_j are empty. This yields $\sigma_1 = \{s_2\}$, $\sigma_3 = \{s_1\}$, $\sigma_4 = \sigma_5 = \{s_2\}$, $\sigma_{N+1} = \{s_q\}$, and all other σ_i are empty. □

Now consider a group G with q elements. Let us identify the group elements with the vertices of Θ_q, i.e., $G = \{s_1, \ldots, s_q\}$. Then G acts not only on Θ_q, but on $(\Theta_q)^{*(N+1)}$ componentwise, as well as on $(2^{[N+1]})_\Delta^{*q}$ if we label the q copies of $2^{[N+1]}$ in the join by $s_1, \ldots, s_q$. In Fig. 4.6, this corresponds to permuting the rows as induced by the group action of G on itself.

Corollary 4.21. *If G is a group with q elements considered as a 0-dimensional simplicial complex, then $E_N G$, i.e., the $(N + 1)$-fold geometric join of G, is equivariantly simplicially homeomorphic to $|(2^{[N+1]})_\Delta^{*q}|$.* □

Now note that $|(2^{[N+1]})_\Delta^{*q}|$ is a subcomplex of $|(2^{[N+1]})^{*q}|$, which in turn is simplicially homeomorphic to the geometric join $(\sigma^N)^{*q}$. Hence we have the following sequence of G-equivariant maps:

$$\iota : |(2^{[N+1]})_\Delta^{*q}| \hookrightarrow |(2^{[N+1]})^{*q}| \xrightarrow{\cong} (\sigma^N)^{*q}.$$

In order to simplify notation, we will identify the elements $x \in |(2^{[N+1]})_\Delta^{*q}|$ with their images $\iota(x)$, i.e., with a vector $\iota(x) = (t_1 x_1, \ldots, t_q x_q)$, where $\sum_{i=1}^q t_i = 1$, and there exist pairwise disjoint $\tau_1, \ldots, \tau_q \leq \sigma^N$ such that $x_1 \in \tau_1, \ldots, x_q \in \tau_q$.

The Continuous Tverberg Theorem in the Prime-Power Case

We will now give a proof of Theorem 4.19.

Proof. The plan is to construct a G-equivariant map $E_N G \to \mathbb{E}$, where $G = (\mathbb{Z}_p)^r$, $\mathbb{E}$ is an N-dimensional real vector space with norm-preserving G-action, and $\mathbb{E}^G = \{x \in \mathbb{E} : gx = x \text{ for all } g \in G\} = \{0\}$, such that any zero of this map yields a desired set of q faces $\tau_1, \ldots, \tau_q \leq \sigma^N$, and then to apply Theorem 1.17.

Consider the d-dimensional affine subspace $\mathbb{A}^d$ of vectors in $\mathbb{R}^{d+1}$ with coordinate sum equal to 1, i.e., $\mathbb{A}^d = \{(x_1, \ldots, x_{d+1})^t \in \mathbb{R}^{d+1} : \sum_{i=1}^{d+1} x_i = 1\}$. Without loss of generality we may assume that $f : \sigma^N \to \mathbb{A}^d$. Consider the following map $\bar{f}$ induced by f:

$$\bar{f} : |(2^{[N+1]})^{*q}_\Delta| \longrightarrow \mathbb{R}^{(d+1)\times q},$$

$$(t_1 x_1, \ldots, t_q x_q) \longmapsto \left(t_1 f(x_1) | \cdots | t_q f(x_q) \right),$$

which is equivariant with respect to the action of the symmetric group $\mathrm{Sym}(q)$ on the coordinates of the join, respectively on the columns of the matrices in $\mathbb{R}^{(d+1)\times q}$. If we identify $G = (\mathbb{Z}_p)^r$ with a subgroup of $\mathrm{Sym}(q)$ via an arbitrary bijection $G \to [q]$, then clearly $\bar{f}$ is also G-equivariant.

We now want to understand the behavior of $\bar{f}$ with respect to the sets of desired faces. Assume that there exist pairwise disjoint simplices $\tau_1, \ldots, \tau_q \leq \sigma^N$ and $x_1 \in \tau_1, \ldots, x_q \in \tau_q$ such that $f(x_1) = \cdots = f(x_q)$ is a point in the intersection $f(\tau_1) \cap \cdots \cap f(\tau_q)$. Then $\bar{f}(\frac{1}{q}x_1, \ldots, \frac{1}{q}x_q) = \left(\frac{1}{q}f(x_1) | \cdots | \frac{1}{q}f(x_q) \right)$ yields a matrix with constant rows. We claim that the converse is also true. Assume that $t_1 f(x_1) = \cdots = t_q f(x_q)$, where $\sum_{i=1}^{q} t_i = 1$ and $x_1 \in \tau_1, \ldots, x_q \in \tau_q$ for some pairwise disjoint $\tau_1, \ldots, \tau_q \leq \sigma^N$. Since $f(x_i) \in \mathbb{A}^d$, the sum over all $d+1$ coordinates of $t_i f(x_i)$ yields t_i, and hence $t_1 = \cdots = t_q$. Therefore $f(x_1) = \cdots = f(x_q)$.

Since we are not interested in the actual values of the constant rows, we may consider a suitable projection. Let $\mathbb{E}$ be the subspace in $\mathbb{R}^{(d+1)\times q}$ of all matrices whose row sums are equal to zero, i.e.,

$$\mathbb{E} = \left\{ (a_{ij}) \in \mathbb{R}^{(d+1)\times q} : \sum_{j=1}^{q} a_{ij} = 0 \text{ for all } i = 1, \ldots, d+1 \right\}.$$

In other words, $\mathbb{E}$ is the orthogonal complement of all matrices with constant rows. As before, $\mathrm{Sym}(q)$, and hence G, acts on $\mathbb{E}$ (in a norm-preserving manner) by permuting columns. Let $\pi : \mathbb{R}^{(d+1)\times q} \to \mathbb{E}$ denote the orthogonal projection. This projection is equivariant with respect to the action of $\mathrm{Sym}(q)$ and hence also with respect to G. Moreover, $\dim \mathbb{E} = (d+1)q - (d+1) = (d+1)(q-1) = N$, and it is easy to see that $\mathbb{E}^G = \{0\}$ is the zero matrix, since G acts transitively on the columns.

By our previous observations, the G-equivariant map

$$\pi \circ \bar{f} : |(2^{[N+1]})_\Delta^{*q}| \longrightarrow \mathbb{E}$$

has the property that its zeros correspond to the desired sets of faces $\tau_1, \ldots, \tau_q \leq \sigma^N$. In order to see this, note that π maps matrices with constant rows to the zero matrix. Now, since $|(2^{[N+1]})_\Delta^{*q}|$ is equivariantly simplicially homeomorphic to $E_N G$, $\pi \circ \bar{f}$ has a zero by Theorem 1.17.   □

4.6   An Obstruction to Graph Planarity

In this section we want to find a criterion to decide, by some easy linear algebra computation, whether a given graph is planar. Our approach will be the following. Given a graph G, define an element $\mathfrak{o}(G)$ in some (cohomology) group with the property that G is planar if and only if $\mathfrak{o}(G)$ is trivial, i.e., the neutral element of the group.

In some sense, we are trying to decide this by analyzing all possible drawings of the graph G at once. This analysis is based on the pairwise crossing of edges. It will turn out that the information about the parity of the number of crossings of independent edges suffices to perform this analysis.

The basic concepts of obstruction theory that we are introducing in this section are valid in much broader generality, cf. [vK32], and this section may be seen as a gentle introduction to these concepts. This section is mostly inspired by the two articles [Tut70] and [Sar91].

The Deleted Product of a Graph

We are going to employ the concept of an abstract deleted product of a graph G, which can be defined more generally for simplicial complexes. Let $G = (V, E)$ be a graph, where we assume as usual that $E \subseteq \binom{V}{2}$ and $V \cap E = \emptyset$. Define the (abstract) deleted product of G by

$$G^\times = \left\{ (\sigma, \tau) : \sigma, \tau \in \binom{V}{1} \dot\cup E, \sigma \cap \tau = \emptyset \right\},$$

i.e., pairs of distinct vertices, pairs consisting of a vertex and a nonincident edge, and pairs of independent edges. In the sequel we will abbreviate the singletons $\{v\} \in \binom{V}{1}$ simply by v and the edges $\{u, v\} \in E$ by uv.

If we think of G as being a simplicial complex, then we have a dimension function

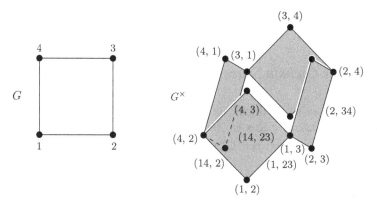

Fig. 4.7 The deleted product of C_4

$$\dim \sigma = \begin{cases} 0, & \text{if } \sigma \in \binom{V}{1}, \\ 1, & \text{if } \sigma \in E, \end{cases}$$

which we extend to $G^\times$ by $\dim(\sigma, \tau) = \dim \sigma + \dim \tau$. If we think of the pair (σ, τ) as the geometric product of σ and τ, then we obtain either a vertex, an edge, or a square. For an illustration of a deleted product, have a look at Fig. 4.7.

Symmetric Cohomology

There is a free $\mathbb{Z}_2$-action on the deleted product $G^\times$ defined by $(\sigma, \tau) \mapsto (\tau, \sigma)$. In order to define the symmetric (respectively, equivariant) cohomology of $G^\times$, consider the *chain groups* $C_i(G^\times)$ defined as the $\mathbb{Z}_2$-vector spaces generated by the elements (cells) of $G^\times$ of dimension i. There are *boundary operators* $\partial : C_i(G^\times) \to C_{i-1}(G^\times)$ defined by $\partial(uv, x) = (u, x) + (v, x)$, $\partial(u, xy) = (u, x) + (u, y)$, and $\partial(uv, xy) = (u, xy) + (v, xy) + (uv, x) + (uv, y)$.

The *symmetric cochains* $C_{\mathbb{Z}_2}^i(G^\times)$ are defined as the sets of linear maps that are invariant with respect to the $\mathbb{Z}_2$-action

$$C_{\mathbb{Z}_2}^i(G^\times) = \{\varphi : C_i(G^\times) \to \mathbb{Z}_2 : \varphi(\sigma, \tau) = \varphi(\tau, \sigma) \text{ for all } (\sigma, \tau) \in G^\times \text{ of dim } i\}.$$

In other words, if we consider the trivial $\mathbb{Z}_2$-action on the coefficient group $\mathbb{Z}_2$, the symmetric cochains are given by equivariant maps.

Coboundary operators $\delta : C_{\mathbb{Z}_2}^i(G^\times) \to C_{\mathbb{Z}_2}^{i+1}(G^\times)$ are defined, as usual, by $\delta(\varphi) = \varphi \circ \partial$. An element in the image of δ will be called a *coboundary*. The obstruction class, $o(G)$, we are going to construct will be an element of the *second symmetric cohomology*

$$H^2_{\mathbb{Z}_2}(G^\times) = C^2_{\mathbb{Z}_2}(G^\times)/\delta C^1_{\mathbb{Z}_2}(G^\times).$$

Bases for the Cochain Spaces

For later purposes let us briefly fix some notation for some standard bases of the symmetric cochain spaces $C^1_{\mathbb{Z}_2}(G^\times)$ and $C^2_{\mathbb{Z}_2}(G^\times)$. First of all, consider the characteristic cochains $[uv, x] \in C^1_{\mathbb{Z}_2}(G^\times)$ and $[uv, xy] \in C^2_{\mathbb{Z}_2}(G^\times)$ defined by

$$[uv, x](\sigma, \tau) = \begin{cases} 1, & \text{if } \sigma = uv \text{ and } \tau = x, \text{ or } \sigma = x \text{ and } \tau = uv, \\ 0, & \text{otherwise,} \end{cases}$$

and

$$[uv, xy](\sigma, \tau) = \begin{cases} 1, & \text{if } \sigma = uv \text{ and } \tau = xy, \text{ or } \sigma = xy \text{ and } \tau = uv, \\ 0, & \text{otherwise.} \end{cases}$$

Since any cochain is constant on $\mathbb{Z}_2$-orbits, it may clearly be written as a sum of characteristic cochains. A basis for $C^1_{\mathbb{Z}_2}(G^\times)$ is given by all characteristic cochains $[uv, x]$, where uv is an edge of G nonincident to x. A collection of characteristic cochains is a basis for $C^2_{\mathbb{Z}_2}(G^\times)$ if it contains, for each *unordered pair* $\{uv, xy\}$ of independent edges, precisely one of the elements $[uv, xy]$, $[xy, uv]$. For example, if there is a linear order $<$ on the set of vertices, then

$$\{[uv, xy] : uv, xy \text{ independent edges of } G, u < v, x < y, u < x\}$$

constitutes a basis.

The Obstruction Class

Let $f = (x, \alpha)$ be a drawing of the graph G as defined in the appendix on page 154. Then f defines a symmetric cochain $\varphi_f : C_2(G^\times) \to \mathbb{Z}_2$ by setting

$$\varphi_f(\sigma, \tau) = |\operatorname{im} \alpha_\sigma \cap \operatorname{im} \alpha_\tau| \mod 2,$$

for generators $(\sigma, \tau) \in C_2(G^\times)$, i.e., the parity of the number of crossings of the edges σ and τ in the drawing f.

As an example, consider the drawing, f, of the complete graph, K_5, on five vertices $V = [5]$, as shown in Fig. 4.8. In this case, $\varphi_f = [13, 24] + [14, 25] + [14, 35]$.

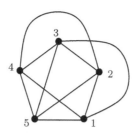

Fig. 4.8 A drawing f of the complete graph K_5

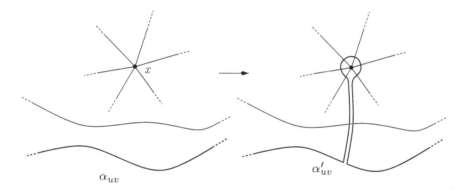

Fig. 4.9 Modification of f via epsilon tubing

We now show that φ_f defines a class in the second symmetric cohomology that depends only on G. In order to show this, it is crucial to understand the image $\delta C^1_{\mathbb{Z}_2}(G^\times) \subseteq C^2_{\mathbb{Z}_2}(G^\times)$. Therefore we are going to compute the coboundaries of the basis elements $[uv, x] \in C^1_{\mathbb{Z}_2}$:

$$\delta[uv, x](\sigma, \tau) = [uv, x](\partial(\sigma, \tau))$$

$$= \begin{cases} 1, & \text{if } \sigma = uv \text{ and } x \in \tau, \text{ or } x \in \sigma \text{ and } \tau = uv, \\ 0, & \text{otherwise.} \end{cases}$$

In other words, $\delta[uv, x]$ evaluates to 1 precisely on the pairs (σ, τ) where σ and τ are both edges, and one is equal to uv, while the other is incident to x.

The addition of such a coboundary, $\delta[uv, x]$, to a cochain, φ_f, can be interpreted geometrically. In fact, there exists a drawing g with the property that $\varphi_g = \varphi_f + \delta[uv, x]$. The drawing g is easily achieved from f by performing an *epsilon tubing* on the edge uv as shown in Fig. 4.9. In other words, the drawing of the edge uv will be changed in such a way that it makes a detour along a small epsilon tube to the vertex x in order to circumscribe a small neighborhood of the vertex x. The course of the tube is such that it will produce only an even number of intersections with

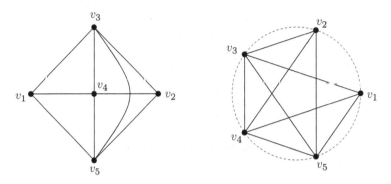

Fig. 4.10 A drawing f and a straight line drawing g of the same graph

other edges along the tube. The circumscription of the vertex x will add 1 to the number of intersections of the new drawing of uv with edges incident to x. Modulo 2, this yields precisely the desired change.

Since the cochains $[uv, x]$ generate $C^1_{\mathbb{Z}_2}(G^\times)$, we obtain the following proposition, which says that every element of the class $[\varphi_f] \in H^2_{\mathbb{Z}_2}(G^\times)$ corresponds to a drawing of the graph G.

Proposition 4.22. *For every drawing f of G and every $\psi \in C^1_{\mathbb{Z}_2}(G^\times)$, there exists a drawing g of G such that $\varphi_g = \varphi_f + \delta\psi$.*   □

It remains to prove the following statement.

Proposition 4.23. *If f and g are drawings of G, then φ_f and φ_g differ by a coboundary, i.e., they define the same cohomology class.*

Proof. Assume that the vertices of G are ordered, $V = \{v_1, \ldots, v_n\}$. By transitivity, it suffices to prove the statement under the additional assumption that the drawing g is a *(standard) straight line drawing*. One way to achieve such a drawing is to evenly distribute the n vertices in counterclockwise order on the unit circle and draw straight edges between adjacent vertices; see, e.g., Fig. 4.10. Triple crossings may be avoided by slightly perturbing the vertices along the circle.

The drawing f may now be modified easily in such a way that the vertices are already in the same position as in the drawing g, and the pairwise intersection of edges does not change at all. For an illustration of such a modification, see Fig. 4.11. Hence, we may assume from now on that f places the vertices in the same position as g.

We now pull the edges in the drawing of f straight one by one and consider the possible changes of the associated cochain. The existence of such a straightening of the edges may be achieved as an application of the Jordan–Schönflies theorem, Theorem A.10. We may assume that each such process results in a drawing of the graph G. Any possible triple intersection or tangential intersection of edges may be removed by a slight perturbation of the edges that haven't been straightened yet.

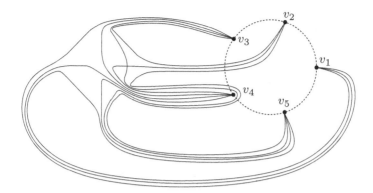

Fig. 4.11 Modification of f to have vertices in the same position as g

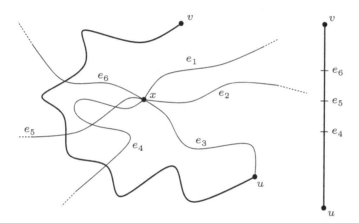

Fig. 4.12 Example edge uv to be straightened, initial position

One representative straightening process is sketched in a process of six steps in Figs. 4.12–4.18. We first consider this particular example and then discuss the general argument. Let's call the initial drawing f_0 and the subsequent drawings $f_1, \ldots, f_6$. In order to analyze the evolution of the associated cochains, $\varphi_{f_0}, \ldots, \varphi_{f_6}$, it clearly suffices to consider the changing intersections of the edge uv with independent edges. In each figure, the right side depicts the intersections of uv with *independent edges*. In other words, intersections with the edge e_3 are not taken into account, since they play no role for the associated cochain.

In step one, from the initial position to position I, a tangential crossing is passed, and this produces a new double intersection with the edge e_4. We clearly have $\varphi_{f_1} = \varphi_{f_0}$. In the step from I to II, the crossing of e_4 and e_5 is passed. The order of intersections of uv with e_4 and e_5 is changed, but $\varphi_{f_2} = \varphi_{f_1}$. The step from II to III works analogously to the previous step, and we have $\varphi_{f_3} = \varphi_{f_2}$. In the step from III to IV, another tangential crossing is passed and a double intersection of uv

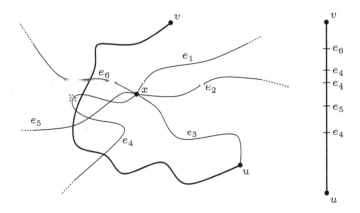

Fig. 4.13 Straightening, position I

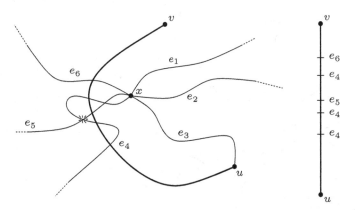

Fig. 4.14 Straightening, position II

with e_4 disappears, and again we have $\varphi_{f_4} = \varphi_{f_3}$. In the step from IV to V, a vertex is passed, the number of intersections with each of the edges e_4, e_5, e_6 decreases by one, and the number of intersections with each of the edges e_1, e_2 (and also e_3) increases by one. In other words, the number of intersections of uv with each edge incident to x changes by one. Hence $\varphi_{f_5} = \varphi_{f_4} + \delta[uv, x]$. In the last step, nothing happens at all, and hence $\varphi_{f_6} = \varphi_{f_5}$. In total, we have $\varphi_{f_6} = \varphi_{f_0} + \delta[uv, x]$.

We now turn to the general argument. In the evolution of intersections of uv with independent edges, the intersection points move along uv. The following phenomena may occur:

1. A tangential crossing is passed: a pair of intersections of uv with some edge e converges and disappears or vice versa.
2. A crossing of e and e' is passed: a pair of intersections of uv with distinct edges e and e' converges and reappears in opposite order.

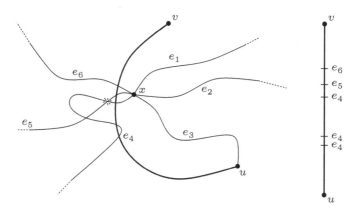

Fig. 4.15 Straightening, position III

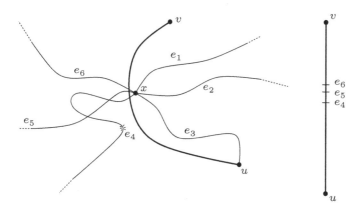

Fig. 4.16 Straightening, position IV

3. A vertex x is passed: a (possibly empty) set of intersections converges, disappears, and a new (possibly empty) set of intersections is born.

In the first two cases, the associated cochain does not change. In the third case, the cochain changes precisely by $\delta[uv, x]$.   □

Summarizing the previous two propositions, we obtain that any two drawings define the same symmetric cohomology class, and every element of this class is the cochain associated to some drawing.

Definition 4.24. Let the *obstruction class*, $\mathfrak{o}(G) \in H^2_{\mathbb{Z}_2}(G^\times)$, be defined by $\mathfrak{o}(G) = [\varphi_f]$, where f is an arbitrary drawing of G.

Now, if G is a planar graph, then clearly the obstruction class, $\mathfrak{o}(G)$, is the zero class, since it contains the cochain associated to a planar drawing. Hence we have a sufficient criterion for nonplanarity of graphs.

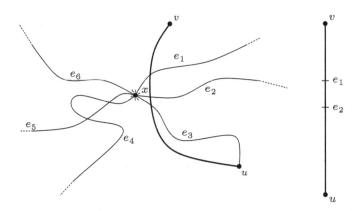

Fig. 4.17 Straightening, position V

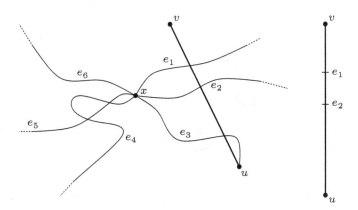

Fig. 4.18 Straightening, position VI

Let us consider the complete graph on five vertices, K_5, as an example. There exists a drawing of K_5 with precisely one pair of edges crossing once and no other pair of edges crossing; see Fig. 4.19. If we denote the associated cochain by φ, then $\varphi = [13, 24]$. Now let $\{i, j, k, l, m\} = [5]$. Then in this particular example, $\delta[ij, k] = [ij, kl] + [ij, km]$, i.e., a sum of two basis elements. In other words, every element of the class $\mathfrak{o}(K_5) = [\varphi]$ can be written only as a sum of an odd number of basis elements. In particular, the class is nonzero, which proves that K_5 is not planar. A similar argument shows that the complete bipartite graph $K_{3,3}$ is not planar.

Remark 4.25. The obstruction classes $\mathfrak{o}(K_5)$ and $\mathfrak{o}(K_{3,3})$ are nonzero.

The justification for the name *obstruction class* is given by the following theorem.

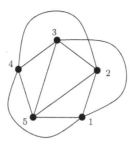

Fig. 4.19 A drawing of the complete graph K_5

Theorem 4.26. *The obstruction class $o(G)$ is zero if and only if the graph G is planar.*

In order to prove this theorem we need two lemmas.

Lemma 4.27. *Let H be a subgraph of the graph G. If $o(H) \neq 0$, then $o(G) \neq 0$.*

Proof. We prove the contrapositive. If $o(G) = 0$, then G has a drawing f such that $\varphi_f = 0$. If we denote the restriction of the drawing f to H by g, then clearly $\varphi_g = 0$. And hence $o(H) = [\varphi_g] = 0$.   □

Recall that a subdivision of a graph is obtained by replacing the edges by paths of length greater than or equal to 1; cf. page 157.

Lemma 4.28. *Let H be a subdivision of the graph K. If $o(K) \neq 0$, then $o(H) \neq 0$.*

Proof. As in the proof of the previous lemma, if $o(H) = 0$, then H has a drawing f such that $\varphi_f = 0$. Use the drawing f in the obvious way to obtain a drawing g of K. Now consider two independent edges e, e' of K. Assume that e is subdivided into a path having edges $e_1, \ldots, e_k$ and e' is subdivided into a path having edges $e_1', \ldots, e_l'$. Then

$$\varphi_g(e, e') = \sum_{i=1}^{k} \sum_{j=1}^{l} \varphi_f(e_i, e_j') = 0.$$

Hence $o(K) = [\varphi_g] = 0$.   □

Proof (of Theorem 4.26). It remains to show that if G is nonplanar, then $o(G)$ is nonzero. But if G is a nonplanar graph, then by Kuratowski's theorem, Theorem A.18, it contains a subgraph H that is a subdivision of either K_5 or $K_{3,3}$. Since both obstruction classes $o(K_5)$ and $o(K_{3,3})$ are nonzero, we obtain by the two previous lemmas that $o(G)$ is nonzero.   □

An immediate corollary of Theorem 4.26 is the following well-known Hanani–Tutte theorem.

Theorem 4.29 (Hanani–Tutte). *If a graph G has a drawing in which any two independent edges intersect an even number of times, then the graph is planar.* $\square$

We could also have used the Hanani–Tutte theorem in order to prove Theorem 4.26. This is interesting, since there are proofs of the Hanani–Tutte theorem that do not make use of Kuratowski's theorem, e.g., [PSŠ07]. In fact, Kuratowski's theorem may, in turn, be obtained from Theorem 4.26, as described in [Sar91].

A Graph Planarity Algorithm

Theorem 4.26 yields an algorithm to determine whether a given graph G is planar. Essentially, the computation boils down to performing some linear algebra over $\mathbb{Z}_2$.

Let a graph $G = (V, E)$ be given. For convenience we assume that $V = [n]$. Fix bases for $C^1_{\mathbb{Z}_2}(G^\times)$ and $C^2_{\mathbb{Z}_2}(G^\times)$ as described before:

$$\mathcal{B}^1 = \{[ab, c] : ab \in E, a < b \text{ and } c \neq a, c \neq b\},$$

$$\mathcal{B}^2 = \{[ij, kl] : ij, kl \in E, i < j, k < l, i < k\}.$$

Compute the matrix representation, $D = (d_{[ij,kl],[ab,c]})$, of the coboundary δ with respect to these bases, i.e.,

$$d_{[ij,kl],[ab,c]} = \begin{cases} 1, & \text{if } ij = ab \text{ and } k = c, \text{ or if } kl = ab \text{ and } i = c, \\ 0, & \text{otherwise.} \end{cases}$$

The cochain φ associated with a standard straight line drawing is easily computed to be

$$\varphi = \sum_{\substack{ij,kl \in E \\ 1 \leq i < k < j < l \leq n}} [ij, kl].$$

Denote by v the vector in $\mathbb{Z}_2^{|\mathcal{B}^2|}$ representing φ with respect to the basis $\mathcal{B}^2$, i.e.,

$$v_{[ij,kl]} = \begin{cases} 1, & \text{if } i < k < j < l, \\ 0, & \text{otherwise.} \end{cases}$$

Now, G is planar if and only if $v \in \text{im}(D : \mathbb{Z}_2^{|\mathcal{B}^1|} \to \mathbb{Z}_2^{|\mathcal{B}^2|})$, which boils down to an easy linear algebra computation.

Before we do an example, it should be mentioned that this algorithm is not very efficient in terms of complexity. There exists an efficient algorithm by John Hopcroft and Robert Tarjan [HT74] that is linear in the number of vertices of the graph. This algorithm was punch-card implemented back in 1974.

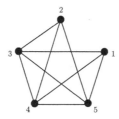

Fig. 4.20 Straight-line drawing of $K_5 \setminus \{e\}$

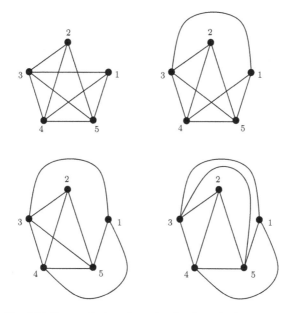

Fig. 4.21 How to obtain a planar drawing from the linear algebra

An Example

Consider the graph $G = K_5 \setminus \{e\}$, i.e., the complete graph on five vertices minus an edge. We use the labeling given in the straight-line drawing shown in Fig. 4.20.

The associated cochain φ is

$$\varphi = [13, 24] + [13, 25] + [14, 25] + [14, 35] + [24, 35].$$

We do not carry out the matrix algebra, but φ clearly is in the image of δ, since for example,

$$\delta([13, 2] + [14, 5] + [35, 2]) = [13, 24] + [13, 25] + [14, 25] + [14, 35] + [24, 35].$$

The element $[13, 2]+[14, 5]+[35, 2]$ is, in fact, telling us how to modify the straight-line drawing in order to obtain a planar drawing of G: pull the edge 13 over the vertex 2, pull 14 over 5, and pull 35 over 2. This is illustrated in Fig. 4.21.

4.7   Conway's Thrackles

We are now going to leave the realm of embeddability questions and turn our attention to other properties of mappings.

This section is about a class of graphs invented by John Conway called *thrackles*. Our presentation is based mostly on the original articles [Woo71, GR92, LPS97, CN00] and the diploma thesis [Bre06].

Let S be a closed surface. A *thrackle drawing* of a graph G in the surface S is a drawing of G in S in which any two adjacent edges do not have any transversal crossings, and any two independent edges have precisely one transversal crossing. A graph G is called *thrackleable (respectively, planar thrackleable)* if there exists a thrackle drawing of G in the 2-sphere (respectively, the plane). We will refer to such drawings as *thrackle drawings* and occasionally to thrackleable graphs simply as *thrackles*.

The three simplest examples are the cycles of lengths 3, 5, and 6 as shown in Fig. 4.22.

Supposedly John Conway explained his knowledge about the word *thrackle* as follows [Arc95]:

> When I was a teenager, on holiday with my parents in Scotland, we once stopped to ask directions of a man who was fishing by the side of a lake. He happened to mention that his line was thrackled. I'd previously called this kind of drawing a tangle, but since I'd just found a knot-theoretical use for that term, I changed this to thrackle. Several people have told me that they've searched in vain for this word in dialect dictionaries, but since I quizzed the fisherman about it, I'm sure I didn't mishear it; he really did use it.

The observant reader has noticed that the cycle of length 4 was missing from the list. And in fact, C_4 is not thrackleable, which easily follows from the Jordan curve theorem. We will show this by means of the concept of thrackle drawings on surfaces on page 125.

Lemma 4.30. *If the cycle C_k is thrackleable for some $k \geq 3$, then C_{k+2} is also thrackleable.*

Proof. This is easily done. In a thrackle drawing of C_k, replace an edge e of the circle by a path of length 3 as shown in Fig. 4.23. In order to do so, find in the

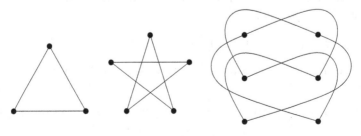

Fig. 4.22 Thrackle drawings of C_3, C_5, and C_6

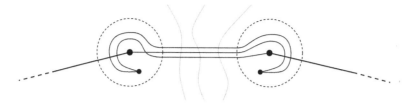

Fig. 4.23 Replacing a single edge by a path of three edges

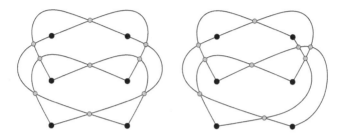

Fig. 4.24 Two cross graphs associated with C_6

original drawing two circles around each vertex of e. If x is a vertex of e, the circle around x has to be so small that it intersects each edge incident to x precisely once. The drawing of the three new edges inside the circles is as depicted in the picture. Outside of the circles the three edges of the path are in a small epsilon neighborhood of the drawing of the original edge, so that the transversal crossings of each new edge are the same as of the original edge.   □

Corollary 4.31. *For $k \neq 4$, all cycles, C_k, are thrackleable.*   □

Another obvious observation is that any subgraph of a thrackleable graph is thrackleable. This will be of some importance in the sequel. In particular, questions about thrackleable graphs maybe reduced to connected graphs.

Lemma 4.32. *If G is thrackleable, then so is every subgraph H of G.*   □

On the other hand, it is immediate that thrackleability is not closed under the graph minor relation (cf. page 157), since clearly C_4 is a topological minor of any larger cycle.

Cross Graphs

Every thrackle drawing determines a planar embedding of an associated cross graph, i.e., the graph that we obtain when placing a vertex at each transversal crossing of edges. This graph may vary depending on the thrackle drawing. Examples of two cross graphs associated with the cycle C_6 are shown in Fig. 4.24.

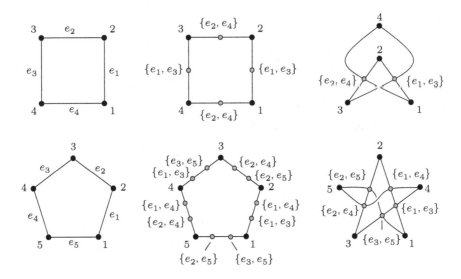

Fig. 4.25 Construction of cross graphs asscociated with C_4 and C_5

We may use this idea to construct thrackle drawings of graphs. To this end, we introduce the general (abstract) concept of a cross graph associated with a graph. A *(thrackle) cross graph G^c associated with a graph $G = (V, E)$* is a graph that is obtained via the following procedure.

For $e \in E$, let i_e be the number of edges in G that are independent of e. Replace each edge e by a path, P_e, of length $i_e + 1$, and arbitrarily label the i_e interior (new) vertices with distinct unordered pairs $\{e, e'\}$, where e' is independent of e.

Observe that for each pair $\{e, e'\}$, there are precisely two vertices with that label, one on P_e and one on $P_{e'}$. The cross graph is now obtained by identifying each pair of such vertices.

Two examples of this construction are shown in Fig. 4.25.

Thrackle Surfaces

In order to produce a drawing of a cross graph G^c, we want to apply the rotation system embeddings explained in Appendix A on page 157. Therefore we have to fix a rotation system.

Choose arbitrary rotations for the original vertices of G. For each new vertex labeled $\{e, e'\}$, choose a rotation in such a way that the incident edges alternate between edges of P_e and $P_{e'}$. We will refer to this condition on the rotation system as the *transversality condition*. Figure 4.26 illustrates two rotations for the vertex labeled $\{e, e'\}$; one satisfies the transversality condition, while the other does not.

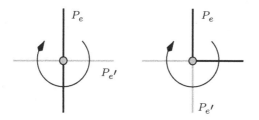

Fig. 4.26 An alternating and a nonalternating rotation

Then each choice of a cross graph G^c together with a rotation system subject to the transversality condition yields an embedding of G^c in some orientable surface M_g that can be lifted to a thrackle drawing of G. This is done by replacing each path P_e with the edge e drawn the same way. Then any two independent edges e and e' will cross transversally precisely once due to the construction of the cross graph and the transversality condition on the rotation of the vertex labeled $\{e, e'\}$ in G^c. Edges e, e' that share a vertex v will not cross, since the paths P_e and $P_{e'}$ share only v as well.

Deciding Thrackleability Combinatorially

Recall that the genus g of a surface can be obtained by purely combinatorial information and is determined by the relation

$$n - m + r = n - m + f = 2 - 2g,$$

where n is the number of vertices, m the number of edges of the graph G^c, and r is the number of orbits of the rotation system, which coincides with the number f of faces of the drawing.

We may apply this relation to decide combinatorially whether a given graph possesses a planar thrackle drawing. For example, in the case of the cycle C_4, there is only one cross graph—shown in Fig. 4.25—and only the alternating rotations at the new vertices have to be fixed, since the original graph vertices have degree 2. It is an easy exercise to see that any choice of a rotation system leads to precisely two orbits of edge walks, one of length 12 and one of length 4. Therefore $n - m + r = 6 - 8 + 2 = 0 = 2 - 2g$ leads to $g = 1$, and hence C_4 has no thrackle drawing on the sphere or in the plane. Figure 4.27 shows a thrackle drawing of C_4 on the torus obtained via a rotation system embedding.

In general, one has to determine the minimum genus, $g_{\min}$, among all possible cross graphs G^c and rotation systems π subject to the transversality condition

$$g_{\min} = \min \frac{1}{2}\left(2 - n(G^c) + m(G^c) - r(G^c, \pi)\right),$$

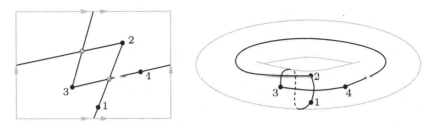

Fig. 4.27 A thrackle drawing of C_4 on the torus

where $n(G^c) = |V(G^c)|$, $m(G^c) = |E(G^c)|$ and $r(G^c, \pi)$ is the number of orbits of the rotation system π.

In summary, we obtain the following statement.

Proposition 4.33. *A graph G is thrackleable if and only if $g_{min} = 0$.*   □

Conway's Thrackle Conjecture

Thrackles have drawn some attention because of a notoriously open conjecture by John Conway about the maximal number of edges a thrackleable graph can have.

Conjecture 4.34 (Conway 1960s). If G is a thrackleable graph with n vertices and m edges, then $m \leq n$.

While the conjecture has not been proven, a variety of beautiful results and insights have been obtained over the decades. On the one hand, we will present a powerful reduction of the problem by Douglas Woodall [Woo71], and on the other we will sketch the ingenious proof of the upper bound $m \leq \frac{3}{2}(n-1)$ by Grant Cairns and Yury Nikolayevsky [CN00] extending previous work by László Lovász, János Pach, and Mario Szegedy [LPS97].

The Eight Graph Version

The fact that thrackleability is closed under the subgraph relation suggests that a forbidden minor approach—as in Kuratowski's theorem characterizing graph planarity—is feasible in order to prove Conway's conjecture. In spirit, this is what we are going to do. We will show that any thrackle drawing of a graph with more edges than vertices yields a drawing of a certain prototype of a graph with precisely one more edge than vertices.

For $i, j \geq 3$, let $8_{i,j}$ be the graph that consists of two cycles, of lengths i and j respectively, that share precisely one vertex; see Fig. 4.28.

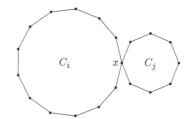

Fig. 4.28 An eight graph $8_{i,j}$ for $i = 13$ and $j = 8$

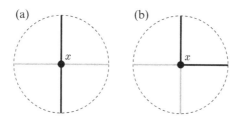

Fig. 4.29 A neighborhood of x in a thrackle drawing of $8_{i,j}$

Theorem 4.35 (Woodall [Woo71]). *The thrackle conjecture is true, provided that for $k \geq 2$, none of the graphs $8_{2k,2k}$ is thrackleable.*

In fact, it follows easily from the proof of Lemma 4.30 that it suffices to require that $8_{2k,2k}$ be not thrackleable for infinitely many $k \geq 2$.

Before we prove Theorem 4.35, we consider how, in a possible thrackle drawing of $8_{i,j}$, the edges of the two cycles meet at the common vertex x. Figure 4.29 illustrates the general picture of an epsilon neighborhood of x (up to homeomorphism); the cycles either (a) cross transversally or (b) touch.

Lemma 4.36. *In a thrackle drawing of $8_{i,j}$, the cycles C_i and C_j cross transversally at the common vertex x if both i and j are odd, and they touch otherwise.*

Proof. Any two circles drawn in the plane such that there are only transversal double intersections cross each other an even number of times. An illustration of this fact is shown in Fig. 4.30, while a proof is left as an exercise.

We now compute the intersections of the drawings of C_i and C_j in a thrackle drawing. Let's first consider the edges of C_i that are not incident to x. Each of these crosses each edge of C_j precisely once. The two edges of C_i incident to x cross all the edges of C_j not incident to x. Hence we obtain

$$(i - 2)j + 2(j - 2) = ij - 4 \equiv ij \quad \mathrm{mod}\ 2.$$

The only possible crossing left to consider occurs at the vertex x. Clearly if i or j is even, then no transversal crossing at x is allowed, but if both i and j are odd, the two cycles have to cross transversally at x. $\qquad \square$

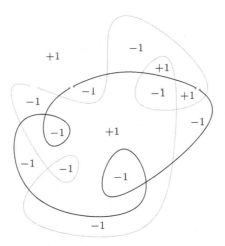

Fig. 4.30 Two *circles* drawn in the plane

A slight generalization of the argument in the previous proof yields the following fact.

Note 4.37. The number of transversal intersections at vertices of any two cycles is congruent modulo 2 to the product of the lengths of the cycles.

This note implies the following interesting corollary, which has important consequences toward transforming a thrackleable graph into a bipartite graph.

Corollary 4.38. *Let G be a thrackleable graph. Then any two odd cycles of G share a vertex.*   □

Proof (of Theorem 4.35). We are going to show that if there exists a thrackle drawing of a graph G with m edges, n vertices, and $m > n$, then there exists a thrackle drawing of an eight graph, $8_{2k,2k}$, for some $k \geq 2$. Assume that G is such a thrackleable graph with $m > n$. Then G also has a connected component with more edges than vertices that is itself thrackleable. Hence we may assume that G is connected. Let T be a spanning subgraph of G. Since T has $n - 1$ edges, there exist at least two more edges e and e' in G. Since the endpoints of each of these two edges determine unique paths in T, the two edges determine unique cycles in G. For an illustration of the three situations that may occur, we refer to Fig. 4.31.

In case (a), the two cycles have precisely one vertex in common; in case (b), the two cycles do not share a vertex, but are connected by a path; and in case (c), the two cycles share a path. We now restrict ourselves to the corresponding subgraphs: in case (a), an eight graph $8_{i,j}$; in case (b), a dumbbell graph; and in case (c), a Θ-graph. We will show that the thrackle drawings in cases (b) and (c) also yield a thrackle drawing of an eight graph, thus leaving us with case (a).

Consider the essential part of the graphs in cases (b) and (c) as illustrated by the dashed box in Fig. 4.32.

(a)   (b)   (c)

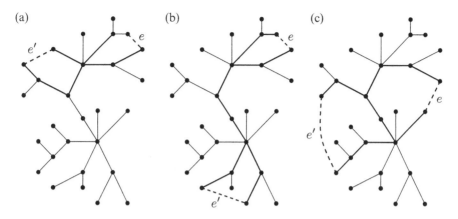

Fig. 4.31 Two cycles determined by the two edges e and e', and their possible interactions

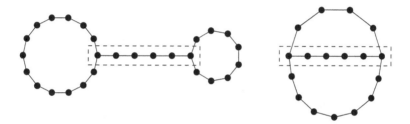

Fig. 4.32 The essential part of a dumbbell graph and of a Θ-graph

We will show how to obtain a planar thrackle drawing of an eight graph from the thrackle drawing of the dumbbell, respectively the Θ-graph, by dividing the path in the essential part into two paths and splitting it at one end. Figure 4.33 shows this procedure for paths of length one, two, and three. The last two cases represent the general case for paths of even and odd length. As before, the pictures are to be understood as follows. No other edges cross the circles, and outside of the circles, the doubled paths continue along the original path in an epsilon neighborhood, so that other edges cross the new doubled edges just as they crossed the original single edges. In order to obtain a clear representation of the situation, we simplify the drawings both here and in the sequel. For example, we do not show the pairwise crossing of the first and the third edges in the path of length three.

Hence we have a thrackle drawing of an eight graph, $8_{i,j}$, for some $i, j \geq 3$. It is left to show that we can obtain a drawing of an eight graph, $8_{2k,2k}$, for some $k \geq 2$. In order to achieve this, we will produce a thrackle drawing of an eight graph, $8_{i',j'}$, with i' and j' even. Then, using the trick from Lemma 4.30 (see Fig. 4.23 on page 123), the length of the shorter of the two cycles may be increased until it matches that of the longer cycle.

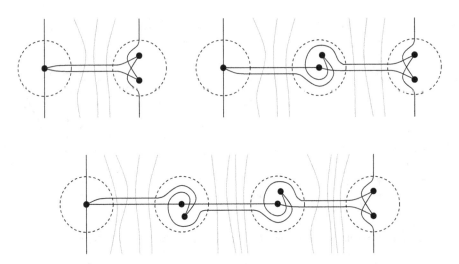

Fig. 4.33 Splitting the path in the essential part

Therefore, if i and j are even, then we are done. Otherwise, either i or j is odd, or both. We will consider separately the cases in which precisely one of them is odd and in which both are odd.

In the first case, we may assume that i is even and j is odd. We will construct a thrackle drawing of $8_{i,2j}$. Lemma 4.36 tells us that the drawings of the two cycles C_i and C_j touch at the common vertex x. We will double the cycle C_j in a two-step procedure depicted in Fig. 4.34. First the path of even length $j - 1$ along C_j is doubled using the doubling of edges as before, and then the crossing at the vertex x is modified as shown in the second step.

In the second case, i and j are odd. We will construct a thrackle drawing of $8_{i+j,j}$ which reduces the problem to the previous case. If both i and j are odd, the drawings of the cycles C_i and C_j cross transversally at the common vertex x according to Lemma 4.36. Now we double a path of odd length $j - 2$ along C_j and then modify the crossing at x in order to obtain a thrackle drawing of $8_{i+j,j}$, as illustrated in Fig. 4.35.   □

Generalized Thrackles

The eight graph version of the thrackle conjecture is a very strong reformulation that suggests new approaches to tackle the conjecture. One immediate thought is to apply the tools of the preceding section using the concept of the obstruction class.

Recall that the obstruction class of a graph G can be represented by the symmetric cochain φ_f asscociated with any drawing f of G, i.e., $\mathfrak{o}(G) = [\varphi_f]$.

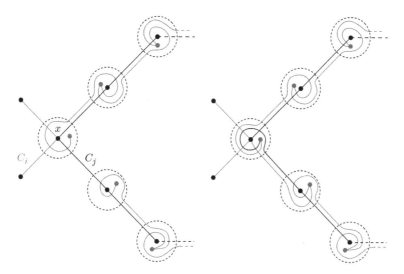

Fig. 4.34 Doubling of the odd cycle C_j in two steps

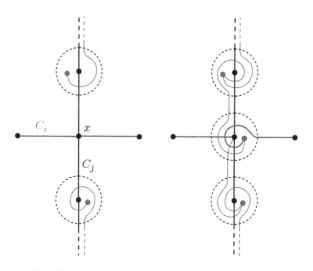

Fig. 4.35 Adding the doubled cycle C_j in two steps to C_i

Now let us denote by φ_k the symmetric cochain $\varphi_k : C_2(8^\times_{2k,2k}) \to \mathbb{Z}_2$ with the property that $\varphi(e, e') = 1$ for each pair of independent edges e, e'. By definition, a thrackle drawing of an eight graph, $8_{2k,2k}$, would have φ_k as associated cochain. Since the eight graphs are planar, we have $\mathfrak{o}(8_{i,j}) = 0$ for all eight graphs $8_{i,j}$. If we could show that $\varphi_k \notin \mathfrak{o}(8^\times_{2k,2k})$, in other words $\varphi_k \notin \text{im}(\delta : C^1_{\mathbb{Z}_2}(8^\times_{2k,2k}) \to C^2_{\mathbb{Z}_2}(8^\times_{2k,2k}))$, then there is no thrackle drawing of $8_{2k,2k}$. Unfortunately, it turns out

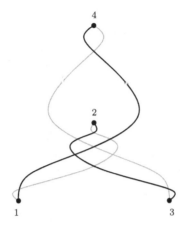

Fig. 4.36 A generalized thrackle drawing of C_4

that $\varphi_k \in \mathfrak{o}(8_{2k,2k})$ for all $k \geq 2$ due to the fact that $8_{2k,2k}$ admits a generalized thrackle drawing, a concept that we will explain now.

Let S be a closed surface. A *generalized thrackle drawing* of a graph G in the surface S is a drawing of G in S in which any two adjacent edges have an even number of transversal crossings, and any two independent edges have an odd number of transversal crossings. A graph G is called *(planar) generalized thrackleable* if there exists a generalized thrackle drawing of G in the sphere, respectively the plane. We will refer to such drawings as *generalized thrackle drawings* and occasionally refer to generalized thrackleable graphs as *generalized thrackles*.

In other words, generalized thrackleability is the mod-2 version of thrackleability with respect to the number of pairwise intersections of edges. Clearly, every thrackleable graph is generalized thrackleable. On the other hand, there are graphs that are generalized thrackles that do not admit a thrackle drawing. Figure 4.36 shows a generalized thrackle drawing of the cycle C_4 that is not thrackleable. For clarity, the edges 12 and 34 are drawn in gray, and the edges 23 and 14 in black.

A consequential result on generalized thrackles is the following characterization in the case of bipartite graphs.

Theorem 4.39. *A bipartite graph G is generalized thrackleable if and only if it is planar.*

Proof (Sketch). Let $G = (V_1 \dot{\cup} V_2, E)$ such that V_1 and V_2 are independent sets of vertices witnessing the bipartiteness of G. We first assume that G is planar. Consider a fixed planar embedding of G and choose a point x in the complement of the drawing. For each vertex $y \in V_1$ choose an x, y-curve that is disjoint from all other vertices and has only transversal crossings with the edges of G. This has to be done in such a way that any two of these curves share only the point x. Consider

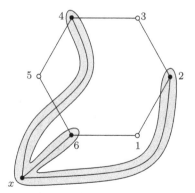

Fig. 4.37 Construction of a disk in a planar drawing of G

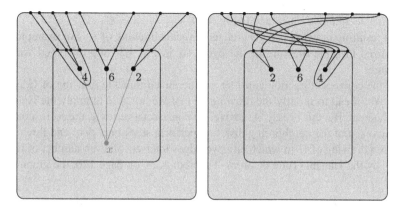

Fig. 4.38 The epsilon disk up to homeomorphism

an epsilon neighborhood of the union of these arcs. We refer to Fig. 4.37 for an illustration.

The left disk of Fig. 4.38 illustrates, up to homeomorphism, the epsilon neighborhood divided into an inner $\frac{1}{2}$-epsilon part and an outer annulus. Note that for each edge $e \in E$, the annulus contains an odd number q_e of line segments belonging to e. In the figure, $q_{56} = 3$ and $q_e = 1$ for all edges $e \neq 56$.

In a *first step*, the drawing of G will be modified within this disk by flipping the inner disk and adjusting the drawing in the annulus accordingly, as illustrated in the right disk of Figure 4.38.

This has the effect that for any two edges $e, e' \in E$, the number of their intersections in the drawing is $q_e q_{e'} \equiv 1 \mod 2$. Moreover, the rotation at each vertex $v \in V_1$ is reversed.

In the *second step*, we will modify the graph so that adjacent edges intersect an even number of times. To this end, consider small epsilon disks about each vertex

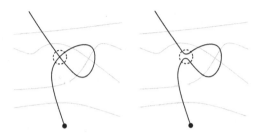

Fig. 4.39 Removing self-intersections

of the graph and perform the above procedure that flips the inner $\frac{1}{2}$-epsilon disk for each of these. This amounts to reversing the rotation at each vertex.

In the *third and last step,* the edges have to be made free of self-intersections by the simple local move illustrated in Fig. 4.39.

The resulting drawing is a generalized thrackle drawing of G. For example, the generalized thrackle drawing of the cycle C_4 in Fig. 4.36 was obtained via this procedure.

For the converse direction consider a generalized thrackle drawing of G in the plane. We intend to modify the drawing so that we are able to apply the Hanani–Tutte theorem. But this is easy. As above, we construct a vertex x, the corresponding x, y-curves, and the neighboring disk and perform steps one, two, and three. The result is a drawing of G in which any two edges intersect an even number of times. Hence, by the Hanani–Tutte theorem, Theorem 4.29, on page 120, the graph G is planar. □

There are some noteworthy details of the proof that we will need later on.

Note 4.40. If $G = (V_1 \dot{\cup} V_2, E)$ is a bipartite graph with a thrackle drawing in the plane, then the drawing can be modified such that

- Every pair of edges has an even number of transversal crossings, and
- The rotation at each vertex $v \in V_1$ is not changed, and the rotation at each vertex $v \in V_2$ is reversed.

An immediate consequence of Theorem 4.39, which we will strongly enhance later on, is the following.

Corollary 4.41. *If G is a thrackleable graph with $n \geq 5$ vertices and m edges, then $m \leq 3n - 6$.*

Proof. By Proposition A.4, there exists a bipartite subgraph G' of G obtained by removing at most half of its edges. Let us denote the number of edges of G' by m'. This bipartite subgraph is thrackleable and therefore does not contain any cycles of length less then 6. If it does not contain any cycles, we know that

$$\frac{m}{2} \leq m' \leq n - 1$$

and hence $m \leq 2(n-1) \leq 3n-6$. If it contains cycles, then

$$\frac{m}{2} \leq m' \leq \frac{3}{2}n - 3,$$

by Theorem A.13, and therefore $m \leq 3n-6$. □

Conway Doubling of an Odd Cycle

Theorem 4.39 is a statement about bipartite graphs. In order to accomplish an upper bound on the number of edges in a thrackleable graph in Corollary 4.41, we used a cheap trick to obtain a bipartite subgraph. This trick didn't take into account that we are in the special situation of a thrackleable graph. The fact that any two odd cycles in a thrackleable graph share a vertex (see Corollary 4.38) can be exploited in a very powerful way.

We will describe a procedure by Conway that doubles an odd cycle in a thrackle drawing of a graph, resulting in a bipartite graph. So let G be a graph with a thrackle drawing and C an odd cycle in G.

For each vertex v of C, consider an appropriate epsilon neighborhood as illustrated in the left drawing of Fig. 4.40. For some chosen orientation of the cycle, the edges incident to vertices of the cycle that are not cycle edges leave the cycle either to the right or to the left. In each epsilon neighborhood, double the vertex v into a *left* and a *right* copy v_l and v_r. Connect those neighbors of v that leave v to the left with v_l and those neighbors that leave v to the right with v_r as shown in the right drawing of Fig. 4.41, where the vertices v_l are depicted in black and the vertices v_r are depicted in white.

Furthermore, for any two neighboring vertices v and v' on C, the new vertices, v_l and v_r', are going to be connected (as will v_r and v_l') by curves in an epsilon neighborhood of the original drawing of the edge vv' as shown in Fig. 4.40. We obtain a thrackle drawing of a new graph $\bar{G}$ in which the cycle C was transformed into a cycle $\bar{C}$ of twice the length. An example of the doubling of a 3-cycle is shown in Fig. 4.41.

Notice that there exists a surjective graph homomorphism $\bar{G} \to G$ that maps each pair of vertices v_l and v_r to the corresponding vertex v and is the identity outside of the cycle.

Note 4.42. If we orient the new cycle according to the orientation of C, then all edges incident to a vertex v_l leave it to the left, and all edges incident to a vertex v_r leave it to the right.

Moreover, some magic has occurred, and all odd cycles have vanished.

Lemma 4.43. *Conway doubling of an odd cycle in a thrackle drawing of a graph produces a bipartite graph.*

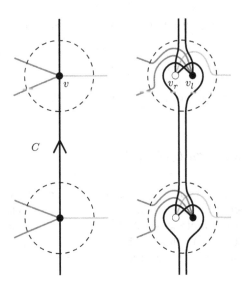

Fig. 4.40 Conway doubling of an edge in a cycle

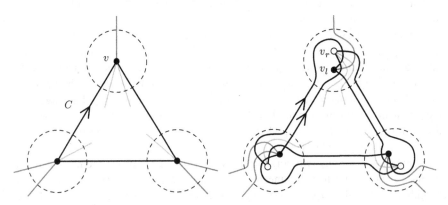

Fig. 4.41 Conway doubling of a 3-cycle

In order to prove this lemma we need to refine Note 4.37 regarding the intersection of cycles in a thrackle drawing. There we observed that the number of transversal intersections at vertices of any two cycles is congruent modulo 2 to the product of the lengths of the cycles.

To refine the note we need a measure for the number of transversal intersections stemming from common vertices and common edges of the cycles. We will define the measure in the generality that is needed later on, i.e., we will consider a closed walk and a cycle. Consider a planar drawing of a graph, G, that may have transversal crossings of edges. Let C_1 be a *closed walk*, i.e., a sequence $v_0, e_1, v_1, e_2, \ldots, v_{k-1}, e_k, v_k = v_0$ such that $e_j = v_{j-1}v_j$, and let C_2 be a cycle in G. For convenience we will consider the indices of the vertices in the walk modulo

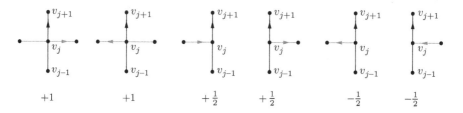

Fig. 4.42 Definition of $\sigma(j)$

k. Provide C_2 with an orientation. For each vertex v_j, $j = 0,\ldots,k-1$, of C_1, define $\sigma(j)$ depending on the orientation of the walk at v_j relative to the orientation of the cycle C_2, as shown in Fig. 4.42. The edges of the walk C_1 are depicted in black, while the edges of the cycle C_2 are shown in gray. In the first two cases, the path v_{j-1}, v_j, v_{j+1} crosses the cycle C_2 transversally; in the other cases, the path v_{j-1}, v_j, v_{j+1} and the cycle C_2 share one edge. In all other cases let $\sigma(j) = 0$.

Define $\sigma(C_1, C_2) = \sum_{j=0}^{k-1} \sigma(j) \mod 2$. This is easily seen to be independent of the orientation of the walk and the cycle.

The idea now is that C_1 and C_2 may be considered continuous drawings of circles in the plane. A slight deformation of each leads to a smooth drawing of two circles with only transversal double crossings. These two curves have an even number of common crossings. Figure 4.43 illustrates this for a drawing of a walk, C_1, and a cycle, C_2, in the plane. The figure also shows the nonzero values of σ at the vertices along C_1.

Observe that the transversal crossings of these two smooth curves are of two different types. The first type is due to the interaction of C_1 and C_2 in the graph and the rotations of common vertices in the drawing, while the second type is due to transversal crossings of edges in the original drawing of the graph G. By construction, $\sigma(C_1, C_2)$ accounts for the number modulo 2 of transversal intersections of C_1 and C_2 of the first type. But since the overall number of transversal crossings of the two curves is even, we conclude that $\sigma(C_1, C_2)$ is congruent modulo 2 to the number of transversal double crossings of edges of the walk C_1 with edges of C_2 in the original drawing of G.

In the very simple example of the figure, there is one transversal double crossing of edges, and hence $\sigma(C_1, C_2) \equiv 1 \mod 2$.

Lemma 4.44. *In a graph G with a thrackle drawing, let C_1 be a closed walk, $v_0, e_1, v_1, e_2, \ldots, v_{k-1}, e_k, v_k = v_0$, and C_2 a cycle. Let $l_1 = k$ be the length of C_1, let l_2 be the length of C_2, and denote by $l(C_1 \cap C_2)$ the number of edges in the walk that are also edges of C_2, i.e., $l(C_1 \cap C_2) = |\{j : e_j \in E(C_2)\}|$. Then the congruence*

$$\sigma(C_1, C_2) \equiv l(C_1 \cap C_2) + l_1 l_2 \mod 2$$

holds.

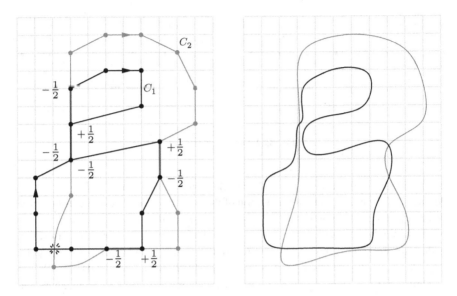

Fig. 4.43 Considering smooth approximating curves with transversal double crossings only

Proof (Sketch). By our previous observations, $\sigma(C_1, C_2)$ is congruent modulo 2 to the number of transversal intersections of edges of the walk C_1 with edges of C_2. To determine this number, we will classify the edges of the walk C_1 with respect to C_2 and consider the corresponding quantities of these.

- Let $k_1 = l(C_1 \cap C_2)$ be the number of edges of the walk C_1 that are also edges of C_2.
- Let k_2 be the number of edges $e_k = v_{k-1}v_k$ of C_1 that are not edges of C_2, but where both v_{k-1} and v_k are vertices of C_2.
- Let k_3 be the number of edges $e_k = v_{k-1}v_k$ of C_1 that are not edges of C_2, but where precisely one of the vertices v_{k-1} or v_k is a vertex of C_2.
- Let k_4 be the number of edges of C_1 that are neither edges of C_2 nor incident to a vertex of C_2.

Since we are considering a thrackle drawing of G, the number of transversal intersections of edges of the walk C_1 with edges of C_2 is

$$k_1(l_2-3) + k_2(l_2-4) + k_3(l_2-2) + k_4 l_2 \equiv k_1 + (k_1+k_2+k_3+k_4)l_2 \quad \text{mod } 2$$

$$\equiv l(C_1 \cap C_2) + l_1 l_2 \quad \text{mod } 2. \qquad \square$$

Equipped with Lemma 4.44, we can now prove that Conway doubling produces a bipartite graph as stated in Lemma 4.43.

Proof (of Lemma 4.43). Suppose that D was the odd cycle on which we performed the Conway doubling in the graph G in order to obtain the graph $\bar{G}$. Denote the doubled cycle in $\bar{G}$ by $\bar{D}$ and let $f : \bar{G} \to G$ be the surjective graph homomorphism that sends $\bar{D}$ to D. Now consider any cycle $\bar{C}$ in $\bar{G}$. The image of $\bar{C}$ under f defines a walk C in G. First of all, outside of the cycles D, respectively $\bar{D}$, the walk C and the cycle $\bar{C}$ agree. Hence, for their respective lengths, we obtain

$$l(C) - l(C \cap D) \equiv l(\bar{C}) - l(\bar{C} \cap \bar{D}) \mod 2.$$

Furthermore by the previous lemma, we have

$$\sigma_G(C, D) \equiv l(C \cap D) + l(C)l(D) \equiv l(C \cap D) + l(C) \mod 2,$$
$$\sigma_{\bar{G}}(\bar{C}, \bar{D}) \equiv l(\bar{C} \cap \bar{D}) + l(\bar{C})l(\bar{D}) \equiv l(\bar{C} \cap \bar{D}) \mod 2.$$

Now it is easy to see that $\sigma_G(C, D) = \sigma_{\bar{G}}(\bar{C}, \bar{D})$, and hence

$$l(\bar{C}) + l(\bar{C} \cap \bar{D}) \equiv l(\bar{C} \cap \bar{D}) \mod 2,$$

with the consequence that $l(\bar{C}) \equiv 0 \mod 2$.   □

Lemma 4.43, together with Theorem 4.39, yields a stronger upper bound on the number of edges in a graph that allows a thrackle drawing than the bound obtained in Corollary 4.41. This is the content of Exercise 17. To obtain the best known bound, we continue our investigation of the properties of Conway's cycle-doubling construction.

The Hanani–Tutte Theorem Revisited

The Hanani–Tutte theorem states that a graph that can be drawn in the plane such that any two independent edges have an even number of transversal crossings is planar. In the proof of Theorem 4.39, we observed that a bipartite graph admitting a thrackle drawing also admits a drawing in which any pair of edges has an even number of transversal crossings. The Hanani–Tutte theorem becomes quite easy to prove under this stronger condition, i.e., that any pair of edges has an even number of transversal crossings. This weak version of the Hanani–Tutte theorem is the content of Theorem A.21. Moreover, the planar drawing can ingeniously be obtained from the original drawing via the following procedure.

First of all, we need the concept of *attaching a handle to an oriented surface*. Let S be a 2-dimensional oriented surface. Consider two disks embedded in the surface. A handle is added along these disks as follows. Cut out the disks, and orient the new boundary circles of the surface according to the orientation of the surface. Now take a cylinder surface, $\mathbb{S}^1 \times [0, 1]$, orient the two boundary circles in an opposite manner, and identify the two boundary circles on the surface with the two boundary circles of

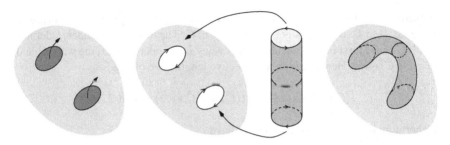

Fig. 4.44 Attaching a handle to a surface

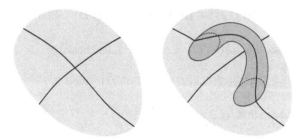

Fig. 4.45 Removing a crossing by attaching a handle

the cylinder in an orientation-preserving way. The result is a new orientable surface with *an attached handle*. For an illustration, see Fig. 4.44.

Now consider a planar drawing of a graph G such that every pair of edges has an even number of transversal crosssings. Eliminate all the crossings by attaching a handle in an epsilon neighborhood of each crossing point, as shown in Fig. 4.45. Note that this does not change the orientations at vertices.

Consider an epsilon neighborhood of the resulting graph embedding on this surface. The boundary components of this epsilon neighborhood are circles. Attach to each boundary component a 2-dimensional disk along the boundary. By Theorem A.21, the graph admits a planar drawing in which the rotations at the vertices are unchanged. Thus, by Theorem A.20, the resulting surface is the 2-sphere. In other words, we have just constructed a planar embedding of the graph G. Figure 4.46 illustrates this procedure in a very simple situation.

Conway Doubling and Planarity

Before we can put the pieces together, we need to make one last observation. Let $G = (V, E)$ be a graph with n vertices and m edges together with a thrackle drawing in the plane. If G is not bipartite, there exists a cycle C in G of length p for some odd $p \geq 3$. The Conway doubling of this cycle produces a bipartite graph $\bar{G} = (V_1 \dot\cup V_2, E')$ with $n + p$ vertices and $m + p$ edges together with a thrackle drawing in

Fig. 4.46 How to obtain a planar embedding

the plane. The graph $\bar{G}$ is planar by Theorem 4.39. In the proof of Theorem 4.39, the drawing of $\bar{G}$ is modified in such a way that any two edges have an even number of transversal crossings. The previous section showed how to obtain a planar drawing of such a graph drawing. The following lemma is essential.

Lemma 4.45. *The doubled cycle bounds a face in the planar embedding of $\bar{G}$.*

Proof. Along the doubled cycle, the *left* vertices, denoted by v_l, and the *right* vertices, denoted by v_r, alternate. Therefore, we may assume that all of the *left* vertices are contained in the independent set V_1, and all of the *right* vertices are contained in V_2. As noted in Note 4.42, the edges leave the vertices in an alternating fashion to the left and to the right along the cycle. As observed in Note 4.40, the modification of the drawing in the proof of Theorem 4.39 leads to a drawing in which each pair of edges crosses an even number of times and reverses the rotations at all vertices of V_2. In other words, if we follow the doubled cycle in the modified drawing of $\bar{G}$, then all edges incident to vertices of the cycle leave the cycle to the left. Hence, in the planar rotation system drawing of this graph, the doubled cycle bounds a face. □

Now we are ready to prove the best known upper bound.

Theorem 4.46. *If G is a graph with n vertices and m edges that admits a thrackle drawing, then $m \leq \frac{3}{2}(n-1)$.*

Proof. We start with the hard case in which G is not bipartite. As in the previous lemma, in the planar embedding of the graph $\bar{G}$, the doubled cycle bounds a face F. Denote the length of F by $l_1 = 2p$ and the lengths of the remaining faces by $l_2, \ldots, l_f$. Since $\bar{G}$ is a bipartite graph with a thrackle drawing, it does not contain odd cycles or 4-cycles. Hence the shortest cycles are of length 6, i.e., $l_2, \ldots, l_f \geq 6$. Hence we obtain the following estimate:

$$2p + 6(f-1) \leq \sum_{j=1}^{f} l_j = 2(m+p).$$

And hence

$$f \le \frac{1}{3}m + 1.$$

Plugging this into Euler's formula (Theorem A.12) yields the desired bound

$$2 = (n + p) - (m + p) + f \le n - m + \frac{1}{3}m + 1.$$

On the other hand, when G is bipartite, we immediately obtain the estimate

$$6f \le 2m,$$

which, together with Euler's formula, leads to the desired upper bound.   □

Exercises

1. Show that the two formulations of the affine Radon theorem, as stated in Theorems 4.1 and 4.2, are equivalent.
2. Provide a proof of the affine Radon theorem (Theorems 4.1 and 4.2) using the fact that any $d + 2$ points in $\mathbb{R}^d$ are *affinely dependent*, i.e., they are not affinely independent as defined on page 174.
3. Show that the map in the proof of Proposition 4.5 is continuous.
4. Provide a proof of Lemma 4.7 on page 100 by applying Lemma 4.8.
5. Provide a proof of Lemma 4.12 on page 102.
6. Provide a proof of Lemma 4.13 on page 103.
7. Consider the Sierksma configuration and the corresponding Tverberg partitions as described on page 106. Show that all Tverberg partitions are obtained in this way and that there are exactly $((q-1)!)^d$ of them.
8. Investigate the Bier sphere $\mathrm{Bier}_n(K)$ for the complex $K = 2^{[n-1]} \subset 2^{[n]}$. Note that $K^\star = K$, and consider the $\mathbb{Z}_2$-action on $\mathrm{Bier}_n(K) = K *_\Delta K$ given by $\sigma * \tau \mapsto \tau * \sigma$. Observe that under the isomorphism defined in the proof of Proposition 4.14, the induced action on the shore subdivision of $\mathrm{Bier}_n(K)$ corresponds to the standard $\mathbb{Z}_2$-action on the barycentric subdivision of the simplex boundary $\mathrm{sd}(2^{[n]} \setminus \{[n]\})$ induced by taking complements. Show that this observation together with Lemma 4.7 yields an alternative proof of Proposition 2.6.
9. (Rectangle peg Problem [Pak08]) Consider a Jordan curve C, i.e., the image of an embedding $f : \mathbb{S}^1 \to \mathbb{R}^2$ of a circle into the plane. Show that there exist four points on the curve C that span a rectangle. Hints: (a) A rectangle is given by four distinct points $a, b, c, d \in C$ such that $\|a - b\| = \|c - d\|$ and $\frac{a+b}{2} = \frac{c+d}{2}$, i.e., the diagonals have the same length and intersect at their midpoints. (b) Consider the configuration space $L = \{\{x, y\} : x, y \in \mathbb{S}^1\}$ of

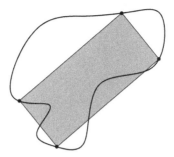

Fig. 4.47 A rectangle on a Jordan curve

two points on the circle, which may coincide, and construct a continuous map $g : L \rightarrow \mathbb{R}^3$ with the property that if it is not injective, then a rectangle on C is found. (c) Consider Exercise 18 on page 196 and recall the definition of the projective plane P. (d) Extend g to a continuous map $\bar{g} : P \rightarrow \mathbb{R}^3$ and recall Proposition 4.16 from page 105 (Fig. 4.47).

10. Show that the Jordan–Schönflies theorem, Theorem A.10, implies that edges can be straightened as proposed on page 114.
11. Explain why—as in the example on page 121—linear algebra gives a recipe for untangling the straight-line drawing of a graph, provided the graph is planar.
12. Give a direct argument that the cycle C_4 of length 4 is not thrackleable.
13. For the unique cross graph of C_4, determine all possible rotation systems subject to the transversality condition and compute the respective orbits of edge walks.
14. Consider a circle drawn smoothly in the plane such that it has only transversal double self-intersections. Show that the connected components of the complement, i.e., the regions of the plane defined by the drawing, may be two-colored in such a way that any two regions sharing a common boundary arc obtain different colors.
15. Give a proof of the fact that any two circles drawn smoothly in the plane with only transversal double crossings cross each other an even number of times, as illustrated in Fig. 4.30.
16. Show that, as needed in the proof of Lemma 4.43, the identity $\sigma_G(C, D) = \sigma_{\bar{G}}(\bar{C}, \bar{D})$ holds.
17. Derive an upper bound for the number of edges of a graph with a thrackle drawing in the plane directly from Lemma 4.43 and Euler's formula.
18. Show that in the proof of Theorem 4.19 as stated on page 109, the fixed-point space $\mathbb{E}^G = \{0\}$ is indeed trivial.

Chapter 5
Appendix A: Basic Concepts from Graph Theory

Graphs serve as a major tool for modeling a wide variety of problems. For any rail system, the plan of stations and possible connections between them constitutes a graph, as do the following: the set of countries on a map, together with the information as to which pairs of those have a common border; the sites of antennas in a wireless communication network, together with the information as to which pairs of those must not use the same frequency in order not to interfere with each other. These, and many other examples, constitute graphs that arise in optimization problems, as well as in other real-world applications. Graphs also play an important role in computer science, pure mathematics, and theoretical physics.

As a general reference I recommend [Die06, Wes05].

A.1 Graphs

A *(finite, simple) graph* G is a pair (V, E), where V is a finite nonempty set, and $E \subseteq \binom{V}{2}$ is a set of 2-element subsets of V. We assume throughout that V is such that $V \cap \binom{V}{2} = \emptyset$, so in particular, $V \cap E = \emptyset$. The elements of V are called *vertices,* and the elements of E are called *edges of G.* If a graph G is given, we refer to its vertex set by $V(G)$ and to its edge set by $E(G)$.

As a first example, let

$$V = [5], E = \{\{1,2\}, \{1,5\}, \{2,3\}, \{2,4\}, \{2,5\}, \{3,5\}\}.$$

Then $G = (V, E)$ can be illustrated as in Fig. A.1. For each vertex of G, a node is drawn, and any two nodes are connected by a line if the corresponding vertices constitute an edge of G.

As long as no confusion can occur, we will usually abbreviate the set $\{u, v\} \in E$ by uv. In the previous example we would therefore write $E = \{12, 15, 23, 24, 25, 35\}$.

M. de Longueville, *A Course in Topological Combinatorics*, Universitext,
DOI 10.1007/978-1-4419-7910-0_5,
© Springer Science+Business Media New York 2013

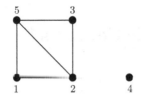

Fig. A.1 A graph on five vertices

If $e = uv \in E$, then we say that u and v are *adjacent,* and that u (resp. v) is *incident* to e. Two distinct edges are *adjacent* if they have a vertex in common. A set of vertices or edges is called *independent* if no two of its elements are adjacent.

We denote by $G \setminus u$ the graph with u removed, i.e., the graph with vertex set $V \setminus \{u\}$ and edge set $E \setminus \{e \in E : u \in e\}$. More generally, for a subset $T \subseteq V$, we consider the graph $G \setminus T = (V \setminus T, \binom{V \setminus T}{2} \cap E)$. Analogously, we denote by $G \setminus e$ the graph with e removed, i.e., the vertex set remains V and the edge set is $E \setminus \{e\}$.

The number of edges incident to a vertex u is called the *degree of u,* which we denote by $d_G(u)$. If no confusion may arise, we also just write $d(u)$.

Lemma A.1. *Let G be a finite simple graph. Then $\sum_{u \in V(G)} d(u) = 2|E(G)|$.*

Informally, this is clear, since every edge has two incident vertices.

Proof (Inductive). We proceed by induction on $m = |E(G)|$. The case $m = 0$ is clear. If $m > 0$, then let $e = vw \in E$ be an arbitrary edge. Then, by the induction hypothesis,

$$2m - 2 = 2(m - 1) = 2|E(G \setminus e)| = \sum_{u \in V} d_{G \setminus e}(u).$$

This, in turn, is equal to

$$\sum_{u \in V \setminus \{v,w\}} d_G(u) + (d_G(v) - 1) + (d_G(w) - 1) = \sum_{u \in V} d_G(u) - 2,$$

from which the result follows.   □

Proof (Double counting). Let $V(G) = \{v_1, \ldots, v_n\}$, $E(G) = \{e_1, \ldots, e_m\}$, and let I be the $n \times m$ *incidence matrix of G* with entries defined by

$$I_{kl} = \begin{cases} 1, & \text{if } v_k \in e_l, \\ 0, & \text{otherwise.} \end{cases}$$

Clearly the number of ones in row k is equal to the number $d(v_k)$ of incident edges, and the number of ones in any column is 2, since every edge has two incident vertices. Now we can count the entries in two different ways, row by row or column by column, to obtain

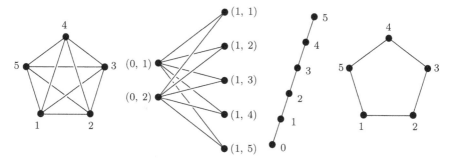

Fig. A.2 Examples of K_n, $K_{m,n}$, P_n, and C_n for $n = 5$ and $m = 2$

$$\sum_{u \in V} d_G(u) = \sum_{k=1}^{n} d_G(v_k) = \sum_{k=1}^{n} \sum_{l=1}^{m} I_{kl}$$

$$= \sum_{l=1}^{m} \sum_{k=1}^{n} I_{kl} = \sum_{l=1}^{m} 2 = 2m,$$

which yields the result.   □

Corollary A.2. *The number of vertices of odd degree in a finite simple graph is even.*   □

Important Graph Classes

Before we proceed, we present some important classes of graphs. The *complete graph,* K_n, is a graph on n vertices in which all pairs of vertices are adjacent. If we define the vertex set of K_n to be $V(K_n) = [n]$, then the edge set is $E(K_n) = \binom{[n]}{2}$. For $m, n \geq 1$, we let $K_{m,n}$ be the *complete bipartite graph* given by the vertex set $V = \{0\} \times [m] \cup \{1\} \times [n]$ and edge set $E = \{uv : u \in \{0\} \times [m], v \in \{1\} \times [n]\}$. In other words, while there are no edges among the vertices of the set $A = \{0\} \times [m]$, nor among the vertices in $B = \{1\} \times [n]$, all possible edges uv with $u \in A$ and $v \in B$ appear. We refer to A and B as the *shores* of the bipartite graph. The *path* P_n *of length* n is defined by the vertex set $V = \{0, 1, \ldots, n\}$ and edge set $E = \{\{i - 1, i\} : i \in [n]\}$. Finally, the *cycle* C_n *of length* n is given by vertex set $V = [n]$ and edge set $E = \{\{i, i + 1\} : i \in [n - 1]\} \cup \{\{1, n\}\}$. For examples, see Fig. A.2.

If $G = (V, E)$ is a graph, then we define its *complement* to be $\bar{G} = (V, \binom{V}{2} \setminus E)$, i.e., the graph on the same vertex set but with complementary edge set. In particular, $\bar{K}_n$ is the graph on vertex set $[n]$ and no edges. Sometimes we will refer to an edge of $\bar{G}$ as a nonedge of G.

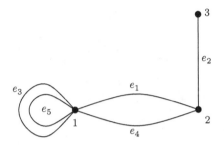

Fig. A.3 An example of a multigraph

Loops

A *graph G with loops* is a pair $G = (V, E)$ with vertex set V and edge set E that consists of 2- and 1-element subsets of V. A 1-subset $\{v\}$ in E is *a loop at v*. We interpret it as an edge with both ends attached to v. In particular, if there exists a loop at v, then v is adjacent to itself. Note that a finite simple graph is a *graph with loops* that does not have loops.

Most of the subsequent concepts work for graphs with loops, though in most cases, we will be thinking of simple graphs.

Multiple Edges

At times it is necessary to consider an even more general graph concept. A *multigraph* is given by a pair $G = (V, E)$ of (finite) disjoint sets of *vertices* and *edges* together with a function $E \to \binom{V}{1} \cup \binom{V}{2}$ that associates the end vertices to each edge. It is clear that every graph (with loops) is a multigraph: the associated function is given by the identity map.

In general, the preimage of an element of $\binom{V}{1} \cup \binom{V}{2}$ may now contain multiple edges. An easy example is given by $V = [3]$ and $E = \{e_1, \ldots, e_5\}$, where the end vertices of the edges are given by

$$e_1 \longmapsto \{1, 2\}, \qquad\qquad e_2 \longmapsto \{2, 3\},$$
$$e_3 \longmapsto \{1\}, \qquad\qquad e_4 \longmapsto \{1, 2\},$$
$$e_5 \longmapsto \{1\}.$$

The corresponding graph is illustrated in Fig. A.3.

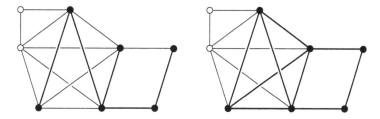

Fig. A.4 A subgraph and an induced subgraph

Homomorphisms

A *graph homomorphism* from G to H is a map $\varphi : V(G) \rightarrow V(H)$ that preserves adjacency, i.e., if $uv \in E(G)$, then $\varphi(u)\varphi(v) \in E(H)$. If φ is a bijection of the vertices such that $uv \in E(G)$ if and only if $\varphi(u)\varphi(v) \in E(H)$, then we call it a *graph isomorphism*. Graph isomorphism yields an equivalence relation on the class of graphs.

Often we will not distinguish between graphs within the same equivalence class. For example, we will call any graph G *complete* if it is isomorphic to the complete graph K_n, where $n = |V(G)|$. We refer to complete bipartite graphs, paths, cycles, etc., in an analogous manner.

Subgraphs

If $G = (V, E)$ is a finite simple graph and if for some $V' \subseteq V$ and $E' \subseteq E$, the pair $G = (V', E')$ is a graph, i.e., $E' \subseteq \binom{V'}{2} \cap E$, then we call G a *subgraph* of G. If $S \subseteq V$ is any subset of the vertices, then we may consider the subgraph of G with vertex set S and all possible edges of G among the vertices of S, i.e., the graph $G[S] = (S, \{uv \in E : u, v \in S\})$. We call $G[S]$ an *induced subgraph* of G. See Fig. A.4 for an illustration.

Note that for $T \subseteq V$, the graph $G \setminus T$ is the induced subgraph $G[V \setminus T]$, whereas $G \setminus e$ is never an induced subgraph for any $e \in E$.

More generally, for any two graphs G and H, we say that G *is an (induced) subgraph of H* if there exist an (induced) subgraph H' of H and a graph isomorphism between G and H'.

Of particular interest are induced subgraphs without edges. They correspond to independent sets of vertices, i.e., $S \subseteq V$ is *an independent set of vertices in G* if the induced subgraph $G[S]$ does not have any edges.

Similarly, we are interested in induced subgraphs that are complete graphs. A set $S \subseteq V$ is called a *clique* if the induced subgraph $G[S]$ is complete.

A *cycle* in a graph is a subgraph isomorphic to a cycle.

Bipartite Graphs

A graph $G = (V, E)$ is called *bipartite* if it is isomorphic to a subgraph of a complete bipartite graph. In other words, the vertices of G may be partitioned into two independent sets V_1 and V_2 with $V = V_1 \dot\cup V_2$.

A cycle of even length is bipartite, since we may distribute the vertices alternatingly into two sets V_1 and V_2. On the other hand, the vertices of an odd cycle cannot be partitioned into two independent sets. Therefore a bipartite graph must not contain a cycle of odd length. The converse is also true and yields a nice characterization of bipartite graphs.

Lemma A.3. *A graph G is bipartite if and only if it does not contain cycles of odd length.*   □

Whenever we write $G = (V_1 \dot\cup V_2, E)$ for a bipartite graph we implicitly assume V_1 and V_2 to be independent sets witnessing the bipartiteness. We will refer to V_1 and V_2 as the *shores* of the graph.

Clearly, every graph can be made bipartite by removing edges, and there is an easy probabilistic argument that no more than half of them need to be removed.

Proposition A.4. *If $G = (V, E)$ is a graph with m edges, then there exists a subset $E' \subset E$ of at most $\frac{m}{2}$ edges such that the graph $(V, E \setminus E')$ is bipartite.*

The following proof will employ the *probabilistic method*, a very powerful concept in graph theory. We will formulate the proof in a rather sloppy but demonstrative language. A good reference is [AS92].

Proof (Sketch). For each vertex v, toss a fair coin in order to assign it to a set A or a set B, each with probability $\frac{1}{2}$. What is the expected number of edges between A and B? For each edge $uv \in E$, each of the following four cases appears with probability $\frac{1}{4}$:

- $u \in A$ and $v \in A$,
- $u \in B$ and $v \in B$,
- $u \in A$ and $v \in B$,
- $u \in B$ and $v \in A$.

Hence, the number of expected edges between A and B is $m(\frac{1}{4} + \frac{1}{4}) = \frac{m}{2}$. Therefore there exists a partition $V = A \dot\cup B$ such that there are at least $\frac{m}{2}$ edges between A and B. It follows that at most $\frac{m}{2}$ edges have to be removed in order to make the sets A and B independent.   □

Basic Constructions

Let $G = (V_0, E_0)$ and $H = (V_1, E_1)$ be two graphs. The ×-*product*, $G \times H$, *of G and H* is defined to have vertex set $V_0 \times V_1$ and edge set $\{\{(u, u'), (v, v')\} : uv \in E_0, u'v' \in E_1\}$. Figure A.5 shows the ×-product of two paths.

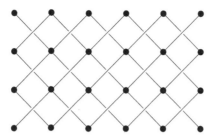

Fig. A.5 The ×-product of P_3 and P_5

The *union of G and H*, $G \cup H$, is the graph with vertex set $V_0 \cup V_1$, and edge set $E_0 \cup E_1$. If $V_0 \cap V_1 \neq \emptyset$, it is often desirable to take the union in such a way that the two individual graphs do not interfere with each other. This is solved by the following type of union.

The *disjoint union of G and H*, $G \amalg H$, is the graph with vertex set $V_0 \times \{0\} \cup V_1 \times \{1\}$, and edge set $\{\{(u, i), (v, i)\} : i = 0, 1, uv \in E_i\}$. The construction makes the vertex sets disjoint by fiat. If $V_0 \cap V_1 = \emptyset$, then $G \amalg H$ is isomorphic to $G \cup H$.

Now let $G = (V, E)$ and let $S \subseteq V$ be an independent set of vertices of G. Denote by G/S the graph in which all vertices of S are identified to one, i.e.,

$$V(G/S) = V \setminus S \dot{\cup} \{s\}$$

for some $s \notin V \setminus S$ and

$$E(G/S) = E \cap \binom{V \setminus S}{2} \cup \{us : \text{there exists } v \in S \text{ such that } uv \in E\}.$$

Connectedness

Every graph $G = (V, E)$ can be decomposed into its connected components in the following way.

A *path from vertex u to v* in a graph G is a sequence P of distinct vertices $v_0 = u, v_1, \ldots, v_k = v$ such that for all i, $v_i v_{i+1} \in E(G)$, i.e., we can walk from u to v along edges in the graph. The *length of P* is defined to be k, i.e., the number of edges along the path. This gives the equivalence relation

$$u \sim v \iff \text{there exists a path from } u \text{ to } v. \tag{A.1}$$

Let $V = V_1 \dot{\cup} \cdots \dot{\cup} V_r$ be the partition of V into equivalence classes. Then the *connected components of G* are defined to be the induced subgraphs $G[V_1], \ldots, G[V_r]$. A graph is *connected* if it has only one connected component.

Trees

A connected graph without cycles is called a *tree*. A vertex of degree one in a tree is called a *leaf*. A *spanning subgraph* of a graph $G = (V, E)$ is a subgraph with vertex set V. A *spanning tree* is a spanning subgraph that is a tree.

Theorem A.5. *If* $G = (V, E)$ *is a graph on* n *vertices, then the following statements are equivalent:*

1. G *is connected and has no cycles,*
2. G *is connected and has* $n - 1$ *edges,*
3. G *has* $n - 1$ *edges and no cycles,*
4. *For any two vertices* $u, v \in V$ *there exists exactly one path from* u *to* v.

$\square$

Corollary A.6. *Every connected graph* G *contains a spanning tree.* $\square$

A *forest* is a graph that has no cycles. Hence, by definition, each connected component of a forest is a tree.

Walks

A slight generalization of a path in a graph is the concept of a walk. A *walk from a vertex* u *to a vertex* v *in a graph* $G = (V, E)$ is an alternating sequence $v_0 = u, e_1, v_1, e_2, \ldots, v_{k-1}, e_k, v_k = v$ of vertices and edges such that $e_i = v_{i-1}v_i$ for all $1 \leq i \leq k$. In other words, a walk may use vertices and edges several times, whereas a path is allowed to visit a vertex of G at most once. The *length* of a walk is defined to be the number of its edges counted with multiplicities, i.e., the length of $v_0, e_1, v_1, e_2, \ldots, v_{k-1}, e_k, v_k$ is k. A walk *closed* if it ends where it starts, i.e., if $v_0 = v_k$.

A.2 Graph Invariants

There exist several important graph parameters that are invariant under graph isomorphisms. We will present several of them.

A *k-coloring* of a graph G is a map $c : V(G) \rightarrow C$ of the vertices to some set of cardinality $|C| = k$. Such a k-coloring is called *proper* if adjacent vertices receive different colors, i.e., $c(u) \neq c(v)$ for all $uv \in E(G)$. The *chromatic number*, $\chi(G)$, of a graph is the smallest number k that allows a proper k-coloring of the graph.

A cycle of even length has chromatic number 2, and a cycle of odd length has chromatic number 3. A complete graph, K_n, has chromatic number n, since any two vertices are adjacent. A bipartite graph has chromatic number 2, one color for each shore.

Two important parameters of G are the *minimal degree, $\delta(G)$*, and the *maximal degree, $\Delta(G)$*, defined to be the minimal, respectively maximal, vertex degree in G.

Cycles have $\delta(G) = \Delta(G) = 2$, while complete graphs have $\delta(K_n) = \Delta(K_n) = n - 1$. A complete bipartite graph $K_{m,n}$ with $m \leq n$ has $\delta(K_{m,n}) = m$ and $\Delta(K_{m,n}) = n$. A graph G satisfying $\delta(G) = \Delta(G) = k$ is called k-*regular*.

It is immediate that for any graph $G = (V, E)$, we obtain an upper bound of $\chi(G) \leq \Delta(G) + 1$ for the chromatic number. Just order the vertices arbitrarily, say $V = \{v_1, \ldots, v_n\}$. Then the *greedy coloring*, c, of G is constructed inductively by first coloring v_1, then v_2, then v_3, and so on, i.e.,

$$c : V \longrightarrow [\Delta(G) + 1],$$

$$v_k \longmapsto \min\left([\Delta(G) + 1] \setminus \{c(v_i) : 1 \leq i < k, v_k v_i \in E\}\right).$$

This procedure can be refined by noticing that to color v_k in the greedy coloring, we need a reservoir of only $d_{G[v_1, \ldots, v_{k-1}]}(v_k) + 1$ colors. In order to minimize this number, choose v_n of minimal degree in G, then choose inductively v_k of minimal degree in $G[\{v_n, \ldots, v_{k+1}\}]$. This yields the bound

$$\chi(G) \leq \max\{\delta(H) : H \text{ an induced subgraph of } G\} + 1, \qquad (A.2)$$

which is the content of Exercise 17. It is in this context that the following classical theorem, which states that the only cases for which $\chi(G) = \Delta(G) + 1$ are the complete graphs and the odd cycles, belongs.

Theorem A.7 (Brooks 1941). *Let G be a connected finite simple graph. If G is neither a complete graph nor a cycle of odd length, then $\chi(G) \leq \Delta(G)$.* □

The *independence number* of G, $\alpha(G)$, is the size of the largest independent set in G, i.e., $\alpha(G) = \max\{|S| : S \subseteq V, S \text{ independent in } G\}$. Since every proper coloring $c : V \to C = \{c_1, \ldots, c_r\}$ gives rise to a partition of the vertex set into independent sets $V = c^{-1}(c_1) \,\dot\cup\, \cdots \cup\, c^{-1}(c_r)$, we obtain the following bound for a coloring with $r = \chi(G)$ colors:

$$|V| = \sum_{i=1}^{r} |c^{-1}(c_i)| \leq r\alpha(G) = \chi(G)\alpha(G),$$

and hence a lower bound on the chromatic number

$$\frac{|V(G)|}{\alpha(G)} \leq \chi(G).$$

The *clique number* of G, $\omega(G)$, is the size of a largest clique in G, i.e., $\omega(G) = \max\{|S| : S \subseteq V, S \text{ is a clique in } G\}$. Clearly, if G has a clique of size k, then $\chi(G) \geq k$, and therefore we obtain another lower bound:

$$\omega(G) \leq \chi(G).$$

A.3   Graph Drawings and Planarity

If we consider the drawings of the example graphs, we realize that there is no way to draw the complete graph K_5 in the plane without having at least one intersection of edges. Compare Figs. A.2 and A.6. But many graphs allow such a drawing, and those give rise to a particularly interesting class of graphs.

We have to make the notion of a graph drawing precise. Let S be a closed surface (as defined on page 172) or the Euclidean plane $\mathbb{R}^2$. A *(simple) curve in* S is the image of a smooth map $\alpha : [0, 1] \to S$ from the unit interval to the surface that is injective in the interior, i.e., $\alpha(s) \neq \alpha(t)$ for all $s, t \in (0, 1)$, $s \neq t$. The points $\alpha(0)$ and $\alpha(1)$ are called the *endpoints of the curve*. For $x, y \in S$, a curve defined by α is called an x, y-*curve* if its endpoints consist of x and y. A *drawing of a graph* $G = (V, E)$ *in* S is given by an assignment $v \mapsto x_v \in S$ of a point for each vertex $v \in V$ and an assignment $e \mapsto \operatorname{im} \alpha_e \subset S$ of a curve for each edge $e \in E$, subject to the following conditions:

- The images of the vertices are distinct, i.e., $x_u \neq x_v$ for $u \neq v$;
- For each $e = uv \in E$ the associated curve α_e is an x_u, x_v-curve;
- For any two edges $e \neq f$, the set of *crossings*, i.e., the set $\{x \in S : \text{there exist } r, s \in (0, 1) \text{ such that } \alpha_e(r) = \alpha_f(s) = x\}$, is finite;
- There are no *triple crossings* of edges, i.e., for any three distinct edges e, f, and g the set $\{x \in S : \text{there exist } r, s, t \in (0, 1) \text{ such that } \alpha_e(r) = \alpha_f(s) = \alpha_g(t) = x\}$ is empty; and
- Crossings are always *transversal* and never *tangential*, i.e., the tangent vectors of two crossing edges in the crossing point always span the tangent plane.

A drawing will be referred to by the pair (x, α). Note that in the case that the surface S is the 2-dimensional sphere $\mathbb{S}^2$, we may use stereographic projection to replace S with the plane $\mathbb{R}^2$ and vice versa.

Definition A.8. A graph G is *planar* if it has a drawing in the sphere (respectively the plane) without crossings. Such a drawing is called a *planar drawing* or *embedding*. A *plane graph* is a graph together with a planar drawing.

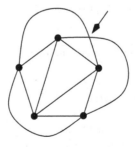

Fig. A.6 Crossing of a pair of edges in a drawing of K_5

In the context of planar drawings, it is "very easy to allow some unproved intuition to creep into the proofs" [Bre93]. The essential theorem to justify most intuition in this context is the Jordan curve theorem [Tho92]. A *Jordan curve* is defined to be a *closed* curve, $c = \mathrm{im}\,\alpha$, in the plane, i.e., a (simple) curve with $\alpha(0) = \alpha(1)$.

Theorem A.9 (Jordan curve theorem). *A Jordan curve partitions the plane into two path-connected components (see page 170) each having c as boundary.*   □

This theorem has a higher-dimensional analogue for embeddings of an $(n - 1)$-dimensional sphere into an n-sphere. But the 2-dimensional case is somewhat special, because the following stronger statement also holds.

Theorem A.10 (Jordan–Schönflies). *Any homeomorphism of two closed curves in the plane can be extended to a homeomorphism of the whole plane. In particular, the closure of the bounded connected component in the complement of a closed curve in the plane is homeomorphic to the 2-dimensional ball $\mathbb{B}^2$.*   □

Euler's Formula

Euler's famous formula establishes a relation between the number of vertices, edges, and faces of a plane graph.

Definition A.11. A *face* of a planar drawing is a path-connected component of the complement of the drawing, i.e., a path-connected component of

$$\mathbb{R}^2 \setminus \bigcup_{e \in E} \mathrm{im}\,\alpha_e.$$

Theorem A.12 (Euler 1758). *Let G be a plane graph with n vertices, m edges, and f faces. Then $n - m + f = 2$.*   □

Consider a connected plane graph G. If we follow the boundary of a face of G, we obtain a closed walk in the graph. Up to the orientation, the walk is uniquely determined by the face. The *length*, $l(F)$, *of a face* F is defined to be the length of one of the closed walks associated with the face.

If G is an arbitrary plane graph, then each face F may have several boundary components each of which determines a walk. The length $l(F)$ is then defined to be the sum of the lengths of all walks in the boundary of F.

Figure A.7 illustrates a face of length 10 and a face of length 8. Clearly, the sum over all lengths of faces yields twice the number of edges in the graph.

Theorem A.13. *Let G be a simple planar graph with n vertices and m edges. Assume that G contains cycles, the shortest of which have length at least k. Then $m \leq \lfloor \frac{k}{k-2}(n - 2) \rfloor$.*

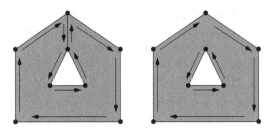

Fig. A.7 Faces of length 10 and 8

Proof. For the sum of the lengths of faces, we obtain $kf \leq \sum_F l(F) = 2m$, and hence by Euler's formula, the desired inequality

$$2 = n - m + f \leq n - m + \frac{2m}{k} = n - \frac{m(k-2)}{k}.$$

□

Corollary A.14. *Every planar graph has a vertex of degree at most 5.*   □

Among planar graphs, the maximal planar graphs constitute an important class. In particular, they are useful in many proofs.

Definition A.15. A planar graph $G = (V, E)$ is called *maximal planar* if it is inclusion-maximal under all planar graphs with vertex set V, i.e., whenever $G' = (V, E')$ is a planar graph with $E \subseteq E'$, then $E = E'$.

Theorem A.16. *For a planar graph G on n vertices, the following are equivalent:*

1. *G has $3n - 6$ edges;*
2. *In every planar drawing in which every edge is represented by a straight line, every face is bounded by a triangle;*
3. *G is maximal planar.*   □

In a maximal planar graph with at least four vertices, the minimal degree is at least three. As in Corollary A.14, it is easy to see that there exist at least four vertices of degree at most 5. Using this fact, it is not hard to show inductively that every planar graph admits a *straight-line embedding*.

Theorem A.17 (Fáry 1948). *Every planar graph admits a drawing in which every edge is represented by a straight line.*   □

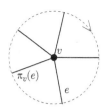

Fig. A.8 Rotation of a vertex

Kuratowski's Theorem

Any graph G' obtained from a graph G by replacing the edges of G by paths of length greater than or equal to 1 is called a *subdivision of G*. A graph H *is a (topological) minor of G* provided that G contains a subgraph that is isomorphic to a subdivision of H. The following is a famous characterization of planar graphs by means of forbidden topological minors.

Theorem A.18 (Kuratowski 1930). *A graph G is planar if and only if neither the complete graph K_5 nor the complete bipartite graph $K_{3,3}$ is a topological minor of G.* ☐

A.4 Rotation Systems and Surface Embeddings

We will now discuss how to obtain graph embeddings in orientable surfaces. A nice overview is given in Carsten Thomassen's handbook article [Tho94]. We recommend that the reader be familiar with the concept of orientable surfaces and 2-cell embeddings as explained in Sect. B.2 on page 172.

If a graph $G = (V, E)$ is 2-cell embedded in an orientable surface S, then the combinatorial essence of the drawing is captured by the walks surrounding the faces. These walks can be described in an alternative way.

Choose an orientation of the surface. For each vertex $v \in E$, the drawing induces a cyclic permutation π_v of the edges incident with v, where $\pi_v(e)$ is defined to be the edge succeeding e in the clockwise orientation around v, as illustrated in Fig. A.8. The permutation π_v is called the *(induced) rotation* of v. The walks surrounding faces can be easily reobtained by the vertex rotations starting with a directed edge and turning clockwise at each vertex v according to the rotation π_v. See Fig. A.9.

Now we want to turn the game around and see how generally to obtain graph embeddings on surfaces.

Definition A.19. Let $G = (V, E)$ be a connected graph. A *rotation system* for G is a family $\{\pi_v : v \in V\}$ of cyclic permutations π_v of the edges incident to v, called *rotations*.

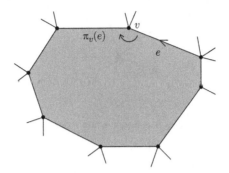

Fig. A.9 Obtaining face walks from vertex rotations

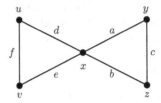

Fig. A.10 Example graph for illustrating orbits of a rotation system

Each directed edge $e_1 = (v_0, v_1)$ gives rise to a sequence $v_0, e_1, v_1, e_2, v_2, \ldots$ with the property that $e_{k+1} = \pi_{v_k}(e_k) = v_k v_{k+1}$. Since the set of directed edges of G is finite, some directed edge will be transversed twice in the same direction. In fact, it is not hard to see that the first directed edge appearing for the second time will be (v_0, v_1). Hence, we obtain a closed walk $v_0, e_1, v_1, \ldots, v_{k-1}, e_k, v_k = v_0$ such that $\pi_{v_k}(e_k) = e_1$ and all directed edges (v_i, v_{i+1}) are distinct. The sequence of directed edges of a walk obtained in such a way (up to cyclic reordering) is called an *orbit* of the rotation system. Clearly every directed edge appears in precisely one orbit. Hence, for every undirected edge, the two directed edges associated with it appear either in the same orbit or in two distinct orbits.

As an example, we consider the graph shown in Fig. A.10. All vertices have degree 2 except for the vertex x. So the only choice that has to be made for a rotation is at this vertex. Consider the two rotations $\pi_x^1 = (abde)$ and $\pi_x^2 = (adbe)$. Up to symmetries of the graph, these are the only two rotations.

The set of closed walks for a rotation system containing π_x^1 consists of

$$x, a, y, c, z, b, x, d, u, f, v, e, x;$$

$$x, b, z, c, y, a, x;$$

$$x, e, v, f, u, d, x;$$

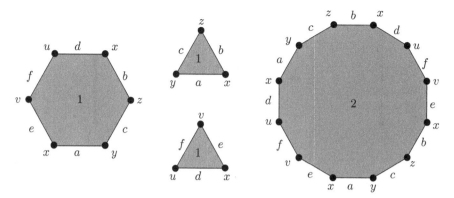

Fig. A.11 The polygons associated with the example rotation systems containing π_x^1, respectively π_x^2

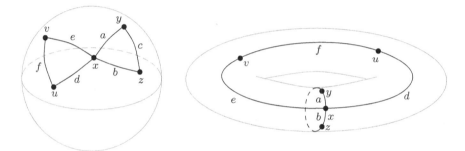

Fig. A.12 G embedded in orientable surfaces arising from rotation systems

and for a rotation system containing π_x^2, consists of the single walk

$$x, a, y, c, z, b, x, e, v, f, u, d, x, b, z, c, y, a, x, d, u, f, v, e, x.$$

For any rotation system of a graph G, we may construct a surface as follows. For each orbit of length k, consider a regular k-gon and identify the directed edges along the polygon with the sequence of directed edges of the orbit. Thereby label the edges of the polygon with the respective edges and the vertices of the polygon with the respective end vertices of the directed edges. The resulting polygons for our two example rotation systems are shown in Fig. A.11.

We now identify each pair of polygon edges with the same label, directed so that the vertex labels coincide as well. This construction results in an orientable surface in which G is embedded. For our two example rotation systems, this yields an embedding of G on the sphere, respectively on the torus, as depicted in Fig. A.12.

Theorem A.20. *Let $\{\pi_v : v \in V\}$ be a rotation system for the connected graph $G = (V, E)$ with n vertices and m edges, and let r be the number of orbits. Then*

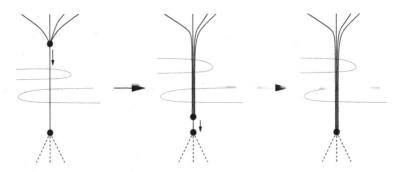

Fig. A.13 Contracting an edge

there exists a 2-cell embedding of G in a pretzel surface of genus g with f faces subject to the relation

$$n - m + f = n - m + r = 2 - 2g.$$ □

A Weak Hanani–Tutte Theorem

It is not hard to extend the notion of rotation systems to multigraphs, i.e., to graphs with loops and multiple edges.

Theorem A.21. *Any multigraph drawn in the plane such that any pair of edges crosses an even number of times can be embedded in the plane without changing the rotation system about the vertices induced by the drawing.*

Proof (Sketch [PSŠ07]). We proceed by induction on the number of vertices. If there is only one vertex with m loop edges, then consider the rotation about this vertex. The parity condition implies that in the rotation, two edges cannot alternate, i.e., two edges e and f cannot appear as $(\cdot\cdot e \cdot\cdot f \cdot\cdot e \cdot\cdot f \cdot\cdot)$. This in turn implies the existence of an edge e that appears as $(\cdots ee \cdots)$ in the rotation. This edge can now be redrawn such that it does not cross any other edges and such that the rotation is not changed. A second induction on the number of edges finishes this case. Now assume that the graph has at least two vertices. Consider a pair of adjacent vertices connected by some edge e. While contracting e, redraw the graph as shown in Fig. A.13.

After this step, some self-intersections of edges might have to be removed as shown in Fig. 4.39 on page 134. By the induction hypothesis, the remaining graph can be embedded in the plane with the same rotation system. Therefore the missing edge can be reintroduced as illustrated in Fig. A.14.   □

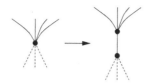

Fig. A.14 Reintroducing the edge

Exercises

1. Show that any two graphs with $n \geq 2$ vertices and $\binom{n}{2} - 1$ edges are isomorphic.
2. Show that every graph G with $n \geq 2$ vertices and $m > \binom{n-1}{2}$ edges is connected.
3. Let u and v be adjacent vertices in a simple graph with n vertices. Show that the edge uv belongs to at least $d(u) + d(v) - n$ triangles in G.
4. Show that among any six people there will always be at least three that pairwise know each other or that pairwise do not know each other.
5. An *automorphism of a graph G* is a graph isomorphism from G to itself. Count the number of automorphisms of P_n, C_n, and K_n.
6. Give a proof of Lemma A.3.
7. Show that the relation (A.1) on page 151 is an equivalence relation.
8. Give a proof of Theorem A.5.
9. Give a proof of Corollary A.6.
10. Let G be a connected simple graph that has neither P_4 nor C_3 as an induced subgraph. Prove that G is a complete bipartite graph.
11. Let T be a tree with $n \geq 2$ vertices. Show that there exist at least two leaves, i.e., two vertices of degree 1.
12. Let G be a forest with n vertices and c connected components. Show that the number of edges is equal to $n - c$.
13. In a graph G, assume that there exists a walk from u to v. Show that there also exists a path from u to v.
14. Let W be a closed walk of length at least one that does not contain a cycle. Show that there exists an edge in the walk that immediately repeats along the walk.
15. Let G be a graph with m edges. Show that the chromatic number $\chi(G)$ satisfies the inequality $\chi(G) \leq \frac{1}{2} + \sqrt{2m + \frac{1}{4}}$.
16. Show that for every graph G, there exists an order of the vertices such that the greedy coloring with respect to this order yields a coloring with $\chi(G)$ colors.
17. Provide the missing details of the proof of the inequality (A.2) on page 153.
18. Let G be a graph in which any two odd cycles share a common vertex. Prove that $\chi(G) \leq 5$.
19. Let G be a graph on n vertices and denote by $\bar{G}$ its complement. Show that $\chi(G) + \chi(\bar{G}) \leq n + 1$.

20. Show that every graph G has a subgraph H with minimal degree $\delta(H) \geq \chi(G) - 1$.
21. Give a proof of Corollary A.14.
22. Use Corollary A.14 to show that every planar graph admits a proper 6-coloring.
23. Provide a proof of Fáry's theorem, Theorem A.17.
24. Let $\{\pi_v : v \in V\}$ be a rotation system for $G = (V, E)$, and (v_0, v_1) a directed edge. Show that (v_0, v_1) is the first directed edge that appears twice in the sequence $v_0, e_1, v_1, e_2, v_2, \ldots$ with the property that $e_{k+1} = \pi_{v_k}(e_k) = v_k v_{k+1}$.
25. Extend the notion of rotation systems to multigraphs.

Chapter 6
Appendix B: Crash Course in Topology

Topology is the mathematical theory of space. It is divided into two main branches: set-theoretic topology and algebraic topology. Set-theoretic topology provides the language of topological spaces and continuous maps, the core concepts such as homotopy and compactness, and properties of some basic constructions. Algebraic topology is essentially a theory about algebraic invariants of topological spaces. It provides the tools to distinguish between spaces with respect to some equivalence relation, primarily homotopy equivalence.

In this overview, the two branches are linked via simplicial complexes and their polyhedra, since these define quite a large class of spaces that allow a convenient way to define one particular class of algebraic invariants, the simplicial homology groups.

For further studies, I recommend the textbooks by Klaus Jänich [Jän84], Glen E. Bredon [Bre93], James Munkres [Mun84] and J. Peter May [May99].

B.1 Some Set-Theoretic Topology

Topological Spaces

Definition B.1. A *topological space* is a pair $(X, \mathcal{O})$, where $X \neq \emptyset$ is a set, and $\mathcal{O} \subseteq \mathcal{P}(X)$ is a family of subsets of X, the family of *open sets*. The set $\mathcal{O}$ is called a *topology* on X and has to satisfy

1. $\emptyset, X \in \mathcal{O}$;
2. Any *finite intersection* $U_1 \cap \cdots \cap U_k$ of open sets $U_1, \ldots, U_k$ is open;
3. Any *arbitrary union* $\bigcup \mathcal{U}$ of open sets $\mathcal{U} \subseteq \mathcal{O}$ is open.

The immediate and extreme examples are the *indiscrete topology* $\{\emptyset, X\}$, and the *discrete topology* $\mathcal{P}(X)$ of all subsets of X. The indiscrete topology is the *coarsest*,

M. de Longueville, *A Course in Topological Combinatorics*, Universitext,
DOI 10.1007/978-1-4419-7910-0_6,
© Springer Science+Business Media New York 2013

and the discrete topology is the *finest,* among all topologies. If $\mathcal{O}_1$ and $\mathcal{O}_2$ are two topologies on X, then $\mathcal{O}_1$ is called *finer* than $\mathcal{O}_2$ if it contains "more" open sets, i.e., $\mathcal{O}_1 \supseteq \mathcal{O}_2$. In this case we say that $\mathcal{O}_2$ is *coarser* than $\mathcal{O}_1$.

All topological spaces that are relevant in this book satisfy the following separation axiom.

Definition B.2. A topological space $(X, \mathcal{O})$ is *Hausdorff* if for any two points $x, y \in X$ with $x \neq y$, there exist open sets $U, V \in \mathcal{O}$ separating x and y, i.e., $x \in U$, $y \in V$ and $U \cap V = \emptyset$.

The most important examples for our purposes are the cases in which X is a subset of Euclidean space $\mathbb{R}^n$ and the topology is induced by the Euclidean metric. We will explain this in more detail now.

Definition B.3. A *metric on a space* X is a map $d : X \times X \longrightarrow \mathbb{R}$ satisfying the following conditions:

1. (Positive definite) $d(x, y) \geq 0$ for all $x, y \in X$, and $d(x, y) = 0$ if and only if $x = y$.
2. (Symmetric) $d(x, y) = d(y, x)$ for all $x, y \in X$.
3. (Triangle inequality) $d(x, z) \leq d(x, y) + d(y, z)$ for all $x, y, z \in X$.

If d is a metric on X and $x \in X$, then denote by $B_\varepsilon(x) = \{y \in X : d(x, y) < \varepsilon\}$ the ε-*ball about* x. The metric d *induces a topology* $\mathcal{O}$ on X: define a set $U \subseteq X$ to be open if for each of its elements $x \in U$ it contains an ε-ball about x for some $\varepsilon > 0$.

Proposition B.4. *If d is a metric on X, and $\mathcal{O}$ is defined by*

$$\mathcal{O} = \{U \subseteq X : \textit{for each } x \in U \textit{ there is an } \varepsilon > 0 \textit{ such that } B_\varepsilon(x) \subseteq U\},$$

then $\mathcal{O}$ is a topology on X.   □

We call such topologies *metric topologies* and the respective spaces *metric topological spaces.* Note that in a metric topological space, the balls $B_\varepsilon(x)$ are open sets, since for each $y \in B_\varepsilon(x)$, the inclusion $B_{\varepsilon - d(x,y)}(y) \subseteq B_\varepsilon(x)$ holds by the triangle inequality.

Proposition B.5. *If d is a metric on X, then the induced topology is Hausdorff.*

Proof. Let $x, y \in X$, $x \neq y$. Then define $\varepsilon = \frac{1}{2} d(x, y)$. The open balls $U = B_\varepsilon(x)$ and $V = B_\varepsilon(y)$ separate x and y as needed.   □

We now introduce the most important terminology for topological spaces.

Definition B.6. Let $(X, \mathcal{O})$ be a topological space and $A \subseteq X$ a subset of X.

- A is *closed* if its complement $X \setminus A$ is open.
- If $x \in X$, then A is called a *neighborhood of x* if there exists an open set $U \in \mathcal{O}$ such that $x \in U \subseteq A$.

- $x \in A$ is an *interior point of A* if A is a neighborhood of x. The set of all interior points of A is denoted by int(A).
- A point $x \in X$ is in the *boundary of A* if it is neither an interior point of A nor an interior point of $X \setminus A$. Equivalently, each neighborhood of x intersects A and $X \setminus A$ nontrivially. The set of all boundary points of A is denoted by ∂A.
- The *closure* of A is defined as cl(A) $= A \cup \partial A$. It is the smallest closed set containing A.
- $x \in X$ is an *accumulation point of A* if $x \in$ cl($A \setminus \{x\}$).

Some of our most frequent examples of topological spaces are the *n-dimensional ball*,

$$\mathbb{B}^n = \{x \in \mathbb{R}^n : \|x\| \leq 1\},$$

and the *$(n-1)$-dimensional sphere*,

$$\mathbb{S}^{n-1} = \{x \in \mathbb{R}^n : \|x\| = 1\},$$

with the topology induced by the Euclidean metric.

Continuous Maps and Homeomorphisms

Definition B.7. Let $(X, \mathcal{O}_X)$ and $(Y, \mathcal{O}_Y)$ be topological spaces. A map $f : X \to Y$ is *continuous* if for any open $V \in \mathcal{O}_Y$, the preimage $f^{-1}(V)$ is open, i.e., $f^{-1}(V) \in \mathcal{O}_X$.

Clearly the identity id$_X$ of a topological space and compositions of continuous maps are continuous.

Note that we will suppress the topology $\mathcal{O}$ if no confusion may arise. Hence we will sometimes speak of spaces X, Y, etc., with the understanding that they are topological spaces $(X, \mathcal{O}_X)$, $(Y, \mathcal{O}_Y)$, etc.

Definition B.8. A map $f : X \to Y$ of topological spaces is a *homeomorphism* if it is continuous, bijective, and the inverse map $f^{-1} : Y \to X$ is continuous as well. Two spaces X and Y are *homeomorphic*, denoted by $X \cong Y$, if there exists a homeomorphism $f : X \to Y$. A map $f : X \to Y$ is called an *embedding* if f is a homeomorphism of X onto its image $f(X)$.

Note that there exist examples of bijective continuous maps $f : X \to Y$ such that the inverse map is not continuous. We leave the construction of an example to the exercises.

Proposition B.9. *Let $(X, \mathcal{O}_X)$ and $(Y, \mathcal{O}_Y)$ be spaces with the topologies $\mathcal{O}_X$ and $\mathcal{O}_Y$ induced by the metrics d_X and d_Y. Then a map $f : X \to Y$ is continuous if and only if for each $x \in X$ and $\varepsilon > 0$, there exists a $\delta > 0$ such that $f(B_\delta(x)) \subseteq B_\varepsilon(f(x))$.* $\square$

Subspaces, Sums, and Products

Proposition B.10. *If* $(X, \mathcal{O})$ *is a topological space and* $Y \subseteq X$ *a subset, then*

$$\mathcal{O}|_Y - \{U \cap Y : U \in \mathcal{O}\}$$

defines a topology on Y, *the* subspace topology. *The space* $(Y, \mathcal{O}|_Y)$ *is called a* subspace *of* $(X, \mathcal{O})$. □

Definition B.11. A *pair of topological spaces* is a pair (X, A), where X is a topological space and A is a subspace. A *map* $f : (X, A) \longrightarrow (X', A')$ *of pairs* is given by a continuous map $f : X \longrightarrow X'$ such that $f(A) \subseteq A'$.

If X and Y are two sets, then their disjoint union, or *sum*, is defined by $X + Y = X \times \{0\} \cup Y \times \{1\}$. If X and Y are disjoint, $X + Y$ may be identified with $X \cup Y$. More generally, if $\{X_\alpha : \alpha \in A\}$ is an indexed family of sets, then their sum is defined by $\coprod_{\alpha \in A} X_\alpha = \bigcup \{X_\alpha \times \{\alpha\} : \alpha \in A\}$. The composition of inclusion maps $X_\alpha \hookrightarrow X_\alpha \times \{\alpha\} \hookrightarrow \coprod_{\alpha \in A} X_\alpha$ is denoted by i_α.

Definition B.12. If $(X, \mathcal{O}_X)$ and $(Y, \mathcal{O}_Y)$ are topological spaces, then their *(topological) sum* is defined to be set $X + Y$ endowed with the topology

$$\{U + V : U \in \mathcal{O}_X, V \in \mathcal{O}_Y\}.$$

More generally, if $\{(X_\alpha, \mathcal{O}_\alpha) : \alpha \in A\}$ is an indexed family of topological spaces, then define the *(topological) sum of the family* to be the set $X = \coprod_{\alpha \in A} X_\alpha$ endowed with the topology

$$\mathcal{O}_X = \left\{ U : U \subseteq \coprod_{\alpha \in A} X_\alpha, i_\alpha^{-1}(U) \in \mathcal{O}_\alpha \right\}.$$

Now, if $\{(X_\alpha, \mathcal{O}_\alpha) : \alpha \in A\}$ is an indexed family of sets, then consider their product $X = \prod_{\alpha \in A} X_\alpha$. If all the X_α are nonempty, then so is X. Note that this statement is not clear a priori but is the content of the *axiom of choice*. For each α there is a projection map $\pi_\alpha : X \longrightarrow X_\alpha$. Finite intersections $\bigcap_{i=1}^k \pi_{\alpha_i}^{-1}(U_{\alpha_i})$, where $\alpha_i \in A$ and $U_{\alpha_i} \in \mathcal{O}_{\alpha_i}$, will be called *basic*.

Definition B.13. Let $\{(X_\alpha, \mathcal{O}_\alpha) : \alpha \in A\}$ be an indexed family of topological spaces. The *(topological) product of the family* is given by the set $X = \prod_{\alpha \in A} X_\alpha$ endowed with the topology

$$\mathcal{O}_X = \left\{ U \subseteq X : U = \bigcup_{\beta \in B} A_\beta \text{ an arbitrary union of basic sets} \right\}.$$

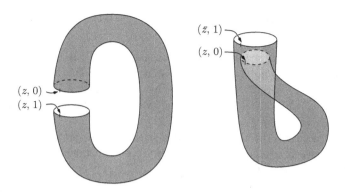

Fig. B.1 The torus and the Klein bottle as quotients of the cylinder

Note that in the case of finitely many spaces $\{X_1, \ldots, X_n\}$ it is common to use the notation

$$X_1 + \cdots + X_n = \coprod_{i=1}^{n} X_i \quad \text{and} \quad X_1 \times \cdots \times X_n = \prod_{i=1}^{n} X_i.$$

Quotients

We want to introduce the concept of a quotient topology. Let $(X, \mathcal{O}_X)$ be a topological space and $q : X \to Y$ a surjective map. Among all topologies on Y, the finest topology such that q is a continuous map is called the *quotient topology with respect to q*. When q is considered as a map of topological spaces, it is called a *quotient map*. The topology is given by

$$\mathcal{O}_Y = \{U \subseteq Y : q^{-1}(U) \text{ is open}\}$$

and has the universal property that a map $g : (Y, \mathcal{O}_Y) \to (Z, \mathcal{O}_Z)$ of topological spaces is continuous if and only if the composition $g \circ q : (X, \mathcal{O}_X) \to (Z, \mathcal{O}_Z)$ is continuous.

A typical example occurs as follows. Consider an equivalence relation $\sim$ on a topological space X. Then there exists a natural (set) quotient map $q : X \to X/\sim$ mapping each point to its equivalence class with respect to $\sim$. The set of equivalence classes $X/\sim$ will now be endowed with the quotient topology with respect to q.

Two concrete examples of such quotients are the *torus* and the *Klein bottle*. Both spaces may be constructed as quotients of the cylinder $S = \mathbb{S}^1 \times [0, 1]$ with respect to an equivalence relation. For the torus, let $\sim_T$ be the equivalence relation generated by $(z, 0) \sim_T (z, 1)$. For the Klein bottle, consider the circle to be a subset of the complex plane, i.e., $\mathbb{S}^1 \subseteq \mathbb{C}$, and consider the equivalence relation generated by $(z, 0) \sim_K (\bar{z}, 1)$. An illustration of the equivalence relations is given in Fig. B.1.

Glued Spaces, Wedges, and Joins

Let X, Y be topological spaces, $A \subseteq X$ a subspace, and $f : A \longrightarrow Y$ a continuous map. Then we may *glue* X to Y along f as follows. The map f defines an equivalence relation $\sim$ on $Y + X$ generated by $f(x) \sim x$ for all $x \in A$. Denote the quotient by $Y \cup_f X = (Y + X)/\sim$ and endow $Y \cup_f X$ with the topology induced by the (set) quotient map $q : Y + X \longrightarrow Y \cup_f X$.

A particular case is the *wedge of pointed spaces*. A *pointed space* is given by a topological space X together with a *base point* $x \in X$, denoted by the pair (X, x). The wedge of the pointed spaces $(X_1, x_1), \ldots, (X_n, x_n)$ is given by gluing together all the base points iteratively. More precisely, let $f_i : \{x_i\} \longrightarrow X_{i+1}$ be defined by $f(x_i) = x_{i+1}$. Then the wedge of the X_i along the base points x_i is iteratively defined by

$$\bigvee_{i=1}^{k+1} X_i = X_{k+1} \cup_{f_i} \left(\bigvee_{i=1}^{k} X_i \right).$$

The wedge of two pointed spaces (X, x) and (Y, y) is also denoted by $X \vee Y$.

Now consider two subspaces $X \subseteq \mathbb{R}^m$ and $Y \subseteq \mathbb{R}^n$. Their join $X * Y$ is defined to be the subspace

$$X * Y = \left\{ \left((1-t)(1, x), t(1, y) \right) \in \mathbb{R}^{m+1} \times \mathbb{R}^{n+1} : t \in [0, 1], x \in X, y \in Y \right\},$$

i.e., the join is essentially the space given by X and Y and all connecting lines between them. Note that for $t = 1$, the element $\left((1-t)(1, x), t(1, y) \right)$ is independent of the choice of x; and for $t = 0$, it is independent of the choice of y. For this reason we suggestively abbreviate the elements of $X * Y$ by pairs $(t_0 x, t_1 y)$, where $t_0, t_1 \geq 0$, $t_0 + t_1 = 1$, and $x \in X$, $y \in Y$. In this notation, we have $(t_0 x, t_1 y) = (t_0' x', t_1' y')$ if and only if the following hold:

- $t_0 = t_0', t_1 = t_1'$, and
- If $t_0 \neq 0$, then $x = x'$, and
- If $t_1 \neq 0$, then $y = y'$.

A special case, which also serves as a good example, is $X = \mathbb{S}^0 = \{\pm 1\} \subseteq \mathbb{R}$ and $Y \subseteq \mathbb{R}^n$ is arbitrary. In this case the join may be realized in dimension $n + 1$, i.e., the map

$$X * Y \longrightarrow \mathbb{R}^{n+1},$$

$$\left((1-t)(1, x), t(1, y) \right) \longmapsto \left((1-t)x, ty \right),$$

is a homeomorphism onto its image, which is easy to check. It turns out that under this map the join $\mathbb{S}^0 * \mathbb{S}^0$ is the boundary of the square $\mathrm{conv}(\pm e_1, \pm e_2)$, $\mathbb{S}^0 * \mathbb{S}^0 * \mathbb{S}^0$

is the boundary of the octahedron $\mathrm{conv}(\pm e_1, \pm e_2, \pm e_3)$, and so forth. In particular, the $(n + 1)$-fold join $\mathbb{S}^0 * \cdots * \mathbb{S}^0$ is the boundary of the $(n + 1)$-*dimensional cross polytope* $Q^{n+1} = \mathrm{conv}(\pm e_1, \ldots, \pm e_{n+1})$, and is homeomorphic to the n-sphere $\mathbb{S}^n$.

A join can also be defined for a pair of arbitrary topological spaces. Let X and Y be topological spaces. Let $\sim$ be the equivalence relation on $X \times [0, 1] \times Y$ generated by $(x, 0, y) \sim (x, 0, y')$ and $(x, 1, y) \sim (x', 1, y)$ for $x, x' \in X$ and $y, y' \in Y$. Then the *join*, $X * Y$, is defined to be the quotient $X \times [0, 1] \times Y / \sim$ equipped with the quotient topology with respect to the (set) quotient map $q : X \times [0, 1] \times Y \to X * Y$.

Again we think of $X * Y$ as being the space of all connecting lines between a point in X and a point in Y. It is shown in Exercise 10 that the general concept of a join and the concept for Euclidean space agree for compact subspaces of Euclidean space.

As in the case for Euclidean space, we will denote an element of an n-fold join $X_1 * \cdots * X_n$ by the formal vector $(t_1 x_1, \ldots, t_n x_n)$, where $(t_1 x_1, \ldots, t_n x_n) = (t_1' x_1', \ldots, t_n' x_n')$ if and only if for all $i = 1, \ldots, n$ we have $t_i = t_i'$ and $x_i = x_i'$ whenever $t_i \neq 0$.

Compactness

A *covering* of a topological space $(X, \mathcal{O})$ is a collection $\mathcal{U}$ of subsets of X whose union is X. A covering is *open* if all sets of the covering are open, i.e., $\mathcal{U} \subseteq \mathcal{O}$. A subset $\mathcal{U}' \subseteq \mathcal{U}$ of a covering is a *subcover* if it is a covering of X itself.

Definition B.14. A topological space X is *compact* if every open covering of X possesses a finite subcover.

The following propositions are easy to derive and are treated in the exercises.

Proposition B.15. *If X is compact and A is a closed subset of X, then A, when viewed as a subspace, is compact.* □

Proposition B.16. *If X is a Hausdorff topological space, then any compact subspace A of X is closed in X.* □

Proposition B.17. *Let X and Y be topological spaces. Then X and Y are compact if and only if the product $X \times Y$ is compact.* □

The *Tychonoff theorem* is a generalization of the previous proposition for arbitrary products of compact spaces.

Proposition B.18. *If X is compact and $f : X \longrightarrow Y$ is continuous, then the image $f(X)$ is compact.* □

Theorem B.19. *If X is compact, Y is Hausdorff, and $f : X \longrightarrow Y$ is a continuous and bijective map, then f is a homeomorphism.* □

Theorem B.20. *If X is a metric space, then the following are equivalent:*

1. *X is compact;*
2. *Each sequence (x_n) in X has an accumulation point; and*
3. *Each sequence (x_n) in X has a convergent subsequence.*   □

Connectivity

Definition B.21. A topological space X is *connected* if it cannot be decomposed into two disjoint open sets.

Equivalently, a space X is connected if the only subsets that are simultaneously open and closed are the empty set $\emptyset$ and the whole space X.

Proposition B.22. *The unit interval $I = [0, 1]$ is connected.*   □

Definition B.23. A topological space X is *path-connected* if for any two points $x, y \in X$, there exists a continuous map $p : [0, 1] \longrightarrow X$ with $p(0) = x$ and $p(1) = y$. The map p is called *a path from x to y.*

Proposition B.24. *A path-connected space X is connected.*   □

Proposition B.25. *If X is (path-)connected, Y is a topological space, and f : $X \to Y$ is a continuous map, then the image $f(X)$ is (path-)connected.*   □

Let X be a topological space. Define a relation $\sim$ on X by

$$x \sim y \iff \text{there exists a path from } x \text{ to } y.$$

The relation $\sim$ is an equivalence relation on X, and the equivalence classes of $\sim$ are the *path-connected components of X*.

Proposition B.26. *Topological spaces X and Y are path-connected if and only if the product $X \times Y$ is path-connected.*   □

Higher Connectivity

A topological space X is *k-connected* if for any $-1 \leq l \leq k$, any continuous map $f : \mathbb{S}^l \to X$ can be extended to the ball $\mathbb{B}^{l+1}$, i.e., there is a map $F : \mathbb{B}^{l+1} \to X$ such that $F|_{\mathbb{S}^l} = f$. An illustration is given in Fig. B.2. For example, a space is -1-connected if it is nonempty; a space is 0-connected if it is nonempty and path-connected; and it is 1-connected (also called *simply connected*) if it is nonempty, path-connected, and every loop can be contracted to a point. If it exists, the largest k such that X is k-connected is called the *connectivity of X*, denoted by $\mathrm{conn}(X)$.

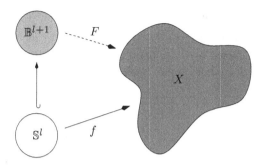

Fig. B.2 Extending a map from the sphere to the ball

An important example is the following theorem. Its proof is the content of Exercises 21–24.

Theorem B.27. *The n-dimensional sphere* $\mathbb{S}^n$ *is* $(n-1)$*-connected.*   □

Homotopy

Let $f, g : X \longrightarrow Y$ be continuous maps of topological spaces, and denote by I the unit interval $[0, 1]$. Then f and g are *homotopic*, denoted by $f \simeq g$, if there exists a *homotopy* between f and g, i.e., a continuous map $h : X \times I \longrightarrow Y$ with $h(x, 0) = f(x)$ and $h(x, 1) = g(x)$ for all $x \in X$.

Proposition B.28. *Homotopy of maps defines an equivalence relation on the space of all continuous maps from X to Y.*   □

A continuous map $f : X \longrightarrow Y$ of topological spaces is a *homotopy equivalence* if there exists a *homotopy inverse* $g : Y \longrightarrow X$, i.e., $g \circ f \simeq \mathrm{id}_X$ and $f \circ g \simeq \mathrm{id}_Y$. Two spaces X and Y are called homotopy equivalent, denoted by $X \simeq Y$, if there exists a homotopy equivalence from X to Y. The class of spaces homotopy equivalent to a space X is called the *homotopy type of X*.

A topological space X is *contractible* if it is homotopy equivalent to a one-point space P. If $f : X \longrightarrow P$ denotes the unique map to a point and $g : P \longrightarrow X$ is the homotopy inverse, then the homotopy h from id_X to $g \circ f$ is called a *contraction* of X.

A subspace $A \subseteq X$ is called a *strong deformation retract* if there exists a continuous *retraction* map $r : X \longrightarrow A$ such that if $i : A \hookrightarrow X$ is the inclusion map, we have $r \circ i = \mathrm{id}_A$ and $i \circ r \simeq \mathrm{id}_X$.

An easy application of Brouwer's fixed-point theorem, Theorem 1.1, yields the following theorem.

Theorem B.29. *The n-dimensional sphere* $\mathbb{S}^n$ *is not a strong deformation retract of the* $(n+1)$*-dimensional ball* $\mathbb{B}^{n+1}$.   □

Corollary B.30. *The n-dimensional sphere $\mathbb{S}^n$ is not n-connected.* $\qquad\qquad\square$

Continuous maps $f, g : (X, A) \longrightarrow (Y, B)$ of pairs of topological spaces are called homotopic, denoted by $f \simeq g$, if $f, g : X \longrightarrow Y$ are homotopic, and $f|_A, g|_A : A \longrightarrow B$ are homotopic. *Homotopy equivalence* of pairs of spaces is defined accordingly.

B.2 Surfaces

A *(closed) surface* S is a compact path-connected Hausdorff space that is *locally homeomorphic* to $\mathbb{R}^2$, i.e., every $x \in S$ possesses an open neighborhood that is homeomorphic to $\mathbb{R}^2$; cf. [Tho94, Mas91]. Typical examples are the 2-dimensional sphere, the torus, and the Klein bottle.

In general, any surface may be obtained as a quotient of one or more polygonal disks whose pairs of directed edges are identified. Some examples are shown in Fig. B.3. Note that the vertices and edges of the polygons yield a graph embedded on the resulting surface. Moreover, the complement of the embedded graph is a disjoint union of open disks. Any embedding of a graph on a surface with the property that the connected components of the complement are open disks is called a 2-*cell embedding*. In this case the connected components of the complement are called the *faces* of the embedding.

A surface that allows a coherent notion of clockwise orientation about every point of the surface is called *orientable*. One way to make this precise for surfaces is to demand that no Möbius strip can be embedded into S. The *Möbius strip* is obtained

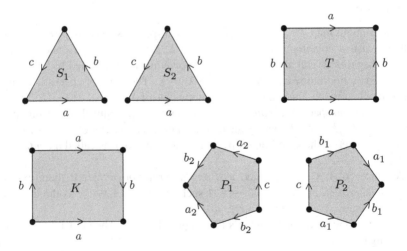

Fig. B.3 The sphere, the torus, the Klein bottle, and a pretzel surface obtained from polygonal disks

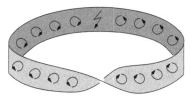

Fig. B.4 A Möbius strip.

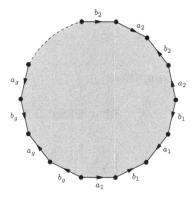

Fig. B.5 A construction recipe for an orientable surface of genus g

as the quotient space $[0, 1] \times [0, 1]/\sim$, where $\sim$ is generated by $(0, t) \sim (1, 1 - t)$. Figure B.4 shows a (stretched) Möbius strip together with an illustration that no coherent orientation is possible. A particular choice of a clockwise orientation for an orientable surface is called an *orientation*.

The most evident examples of nonorientable surfaces are the *projective plane* and the *Klein bottle*. The projective plane may be obtained by gluing a disk to a Möbius strip along their 1-dimensional sphere boundaries. The Klein bottle may be obtained from a polygonal disk as shown in Fig. B.3.

The orientable surfaces are given by the family of pretzel surfaces [Mas77, See06], starting with the 2-dimensional sphere and the torus. In general, a *pretzel surface* is obtained via successively attaching handles to the sphere as described on page 139. The number of handles is called the *genus* of the surface. The pretzel surface of genus g can also be obtained as a quotient of a $4g$-gon, where the edges are identified as shown in Fig. B.5.

Euler Characteristic of Orientable Surfaces

We define the *Euler characteristic* of an orientable surface S of genus g to be $\chi(S) = 2 - 2g$.

Theorem B.31 (Euler–Poincaré). *If a graph G with n vertices and m edges is 2-cell embeddeded with f faces in an orientable surface S of genus g, then the relation*

$$\chi(S) = 2 - 2g = n - m + f$$

holds.   □

B.3   Simplicial Complexes

Geometric Simplicial Complexes

The *affine span* of a set $x_0, \ldots, x_n \in \mathbb{R}^m$ is defined as the smallest affine subspace of $\mathbb{R}^m$ containing $x_0, \ldots, x_n$:

$$\mathrm{aff}(x_0, \ldots, x_n) = \left\{ \sum_{i=0}^{n} \lambda_i x_i : \sum_{i=0}^{n} \lambda_i = 1 \right\}.$$

The points $x_0, \ldots, x_n$ are *affinely independent* if the affine span, $\mathrm{aff}(x_0, \ldots, x_n)$, is n-dimensional. Equivalently, if $v = \sum_{i=0}^{n} \lambda_i x_i \in \mathrm{aff}(x_0, \ldots, x_n)$, then the coefficients $\lambda_0, \ldots, \lambda_n$ are uniquely determined.

An important example is the set of standard basis vectors $e_1, \ldots, e_{n+1} \in \mathbb{R}^{n+1}$. They span the hyperplane in $\mathbb{R}^{n+1}$ of all points with coordinate sum equal to one.

If $x_0, \ldots, x_n \in \mathbb{R}^m$ are affinely independent, then the *(affine) simplex* σ *spanned by* $x_0, \ldots, x_n$ is defined to be the their convex hull

$$\sigma = \mathrm{conv}(x_0, \ldots, x_n) = \left\{ \sum_{i=0}^{n} \lambda_i x_i : \lambda_i \geq 0, \sum_{i=0}^{n} \lambda_i = 1 \right\}.$$

The λ_i are called *barycentric coordinates of* σ. We call $\{x_0, \ldots, x_n\}$ the *vertex set of* σ, denoted by $\mathrm{vert}(\sigma)$. For any k with $-1 \leq k \leq n$, and any $(k+1)$-element subset $\{x_{i_0}, \ldots, x_{i_k}\} \subseteq \{x_0, \ldots, x_n\}$, the simplex $\tau = \mathrm{conv}(x_{i_0}, \ldots, x_{i_k})$ is called a *k-dimensional face of* σ. We denote this by $\tau \leq \sigma$ and $\dim(\tau) = k$. Note that the case $k = -1$ yields that the empty set is a face of σ of dimension -1. The faces of σ of codimension one, i.e., the faces of dimension $n - 1$, will be referred to as *facets of* σ. The *geometric boundary*, $\partial\sigma$, of σ is defined to be the union of all its proper faces, in other words, $\partial\sigma = \bigcup_{\tau < \sigma} \tau$.

The prototype of an n-dimensional simplex is given by the *standard n-simplex* σ^n defined by the standard basis vectors $e_1, \ldots, e_{n+1} \in \mathbb{R}^{n+1}$:

$$\sigma^n = \mathrm{conv}(e_1, \ldots, e_{n+1}) \subseteq \mathbb{R}^{n+1}.$$

Definition B.32. A *(finite geometric) simplicial complex* Δ is a finite family of (affine) simplices in some $\mathbb{R}^m$ such that

1. $\sigma \in \Delta$ and $\tau \leq \sigma$ implies $\tau \in \Delta$;
2. If $\sigma, \tau \in \Delta$, then $\sigma \cap \tau$ is a face of σ and a face of τ.

The simplices $\sigma \in \Delta$ are also referred to as *faces of Δ*, and the inclusion maximal faces of Δ are referred to as *facets of Δ*. The *vertex set of Δ* is defined as the union of the vertices of the faces, $\text{vert}(\Delta) = \bigcup_{\sigma \in \Delta} \text{vert}(\sigma)$. The *dimension of Δ* is defined to be the maximum dimension of a face, $\dim(\Delta) = \max_{\sigma \in \Delta} \dim(\sigma)$. A simplicial complex is called *pure of dimension n* if all its facets have dimension n.

Our first important example of a simplicial complex is given by a simplex together with all its faces. In particular, the standard n-simplex gives rise to the simplicial complex $\Delta^n = \{\tau : \tau \leq \sigma^n\}$.

Even though this seems clear, there is actually a minor lemma to prove.

Lemma B.33. *If $\{x_0, \ldots, x_n\} \subseteq \mathbb{R}^m$ is affinely independent, $A = \{i_0, \ldots, i_k\}$, $B = \{j_0, \ldots, j_l\} \subseteq \{0, \ldots, n\}$ subsets with intersection $C = A \cap B = \{i_0, \ldots, i_m\}$, then*

$$\text{conv}(x_{i_0}, \ldots, x_{i_k}) \cap \text{conv}(x_{j_0}, \ldots, x_{j_l}) = \text{conv}(x_{i_0}, \ldots, x_{i_m}).$$

Proof. Clearly, the right-hand-side simplex is contained in the left-hand-side intersection. Now let

$$v = \sum_{i \in A} \lambda_i x_i = \sum_{i \in B} \mu_i x_i \in \text{conv}(x_{i_0}, \ldots, x_{i_k}) \cap \text{conv}(x_{j_0}, \ldots, x_{j_l}).$$

Then

$$0 = \sum_{i \in A} \lambda_i x_i - \sum_{i \in B} \mu_i x_i = \sum_{i \in A \setminus C} \lambda_i x_i - \sum_{i \in B \setminus C} \mu_i x_i + \sum_{i \in C} (\lambda_i - \mu_i) x_i.$$

As shown in one of the exercises, the affine independence of $\{x_0, \ldots, x_n\}$ implies that all coefficients on the right-hand side have to be zero. In particular, $\lambda_i = 0$ for all $i \in A \setminus C$. And hence

$$v = \sum_{i \in C} \lambda_i x_i \in \text{conv}(x_{i_0}, \ldots, x_{i_m}).$$

$\square$

Another very important example is the boundary complex, Γ^n, of the cross polytope Q^{n+1}:

$$\Gamma^n = \left\{\text{conv}(\varepsilon_{i_0} e_{i_0}, \varepsilon_{i_1} e_{i_1}, \ldots, \varepsilon_{i_k} e_{i_k}) : 1 \leq i_0 < \cdots < i_k \leq n+1, \varepsilon_{i_j} \in \{\pm 1\}\right\}.$$

The space given by the union of all simplices of a simplicial complex Δ is called the *polyhedron of Δ*, denoted by $|\Delta| = \bigcup_{\sigma \in \Delta} \sigma$. Sometimes we refer to $|\Delta|$ when talking about topological properties of Δ.

A topological space X is a *polyhedron* if there exist a simplicial complex Δ and a homeomorphism $h : |\Delta| \to X$. The homeomorphism h together with Δ is called a *triangulation of X*.

Definition B.34. Let Δ and Δ' be simplicial complexes. A *simplicial map from Δ to Δ'* is a continuous map $f : |\Delta| \longrightarrow |\Delta'|$ such that for all $\sigma \in \Delta$:

1. $f(\sigma) \in \Delta'$;
2. The restriction $f|_\sigma : \sigma \longrightarrow f(\sigma)$ is an affine linear map.

Note that the second condition implies that a simplicial map is already determined by its restriction to the vertex set $f|_{\text{vert}(\Delta)}$.

Definition B.35. A *subcomplex* of a simplicial complex Δ is a simplicial complex Γ that is contained in Δ. A *pair of simplicial complexes* is a pair (Δ, Γ), where Δ is a simplicial complex and Γ is a subcomplex. A *map* $f : (\Delta, \Gamma) \longrightarrow (\Delta', \Gamma')$ of *pairs* is given by a simplicial map $f : \Delta \longrightarrow \Delta'$ such that $f(\Gamma) \subseteq \Gamma'$.

Definition B.36. If Δ and Δ' are simplicial complexes, then Δ' is called a *subdivision of Δ* if $\text{vert}(\Delta) \subseteq \text{vert}(\Delta')$ and $|\Delta| = |\Delta'|$.

Abstract Simplicial Complexes

If we restrict our attention to the inclusion relations of the vertex sets of simplices in a simplicial complex, then we obtain the concept of an abstract simplicial complex.

Definition B.37. Let V be a finite set and $K \subseteq \mathcal{P}(V)$ a family of subsets of V. Then K is called an *(abstract) simplicial complex* if it is closed under taking subsets, i.e.,

$$\sigma \in K, \tau \subseteq \sigma \text{ implies } \tau \in K.$$

The elements σ of a simplicial complex K are called *simplices*, while the elements of the simplices are referred to as *vertices*. The *vertex set of K* is the union of all simplices $\text{vert}(K) = \bigcup_{\sigma \in K} \sigma \subseteq V$. A subset $\tau \subseteq \sigma$ is called a *face of σ*, denoted by $\tau \leq \sigma$. The *dimension of a simplex* is defined to be its cardinality minus one, i.e., $\dim(\sigma) = |\sigma| - 1$. A face $\tau \leq \sigma$ with $\dim(\tau) = \dim(\sigma) - 1$ is called a *facet of σ*. In contrast, the *facets of a simplicial complex K* are given by the inclusion maximal simplices. The complex K is called *pure of dimension n* if all its facets are of dimension n. The *dimension of K* is the maximum dimension of a simplex in K, $\dim(K) = \max_{\sigma \in K} \dim(\sigma)$.

Any geometric simplicial complex Δ gives rise to an *associated abstract simplicial complex* $K(\Delta) = \{\text{vert}(\sigma) : \sigma \in K\}$.

Definition B.38. A *simplicial map* of (abstract) simplicial complexes K and L is defined to be a map $f : \text{vert}(K) \longrightarrow \text{vert}(L)$ preserving simplices, i.e., $f(\sigma) \in L$ for each simplex $\sigma \in K$.

Geometric Realizations

We usually think of an abstract simplicial complex in terms of a *geometric realization*.

Definition B.39. A geometric simplicial complex Δ is a *geometric realization* of an abstract simplicial complex K if there exists a bijection $\varphi : \text{vert}(K) \longrightarrow \text{vert}(\Delta)$ such that for any set $v_0, \ldots, v_n \in \text{vert}(K)$ of vertices of K,

$$\{v_0, \ldots, v_n\} \in K \iff \text{conv}(\varphi(v_0), \ldots, \varphi(v_n)) \in \Delta.$$

In other words, Δ is a geometric realization of K if and only if its associated abstract simplicial complex $K(\Delta)$ is simplicially isomorphic to K.

Note that if Δ and Γ are geometric realizations of K and L, respectively, then there is a one-to-one correspondence between simplicial maps $K \to L$ and simplicial maps $\Delta \to \Gamma$.

We will now show that every abstract simplicial complex has a geometric realization.

Lemma B.40. *Any* $m + 1$ *distinct points* $\gamma(t_0), \ldots, \gamma(t_m)$ *on the* moment curve *defined by*

$$\gamma : \mathbb{R} \longrightarrow \mathbb{R}^m,$$

$$t \longmapsto (t, t^2, \ldots, t^m),$$

are affinely independent.

Proof. As shown in the exercises, it suffices to show that the $m + 1$ vectors $(1, t_0, \ldots, t_0^m), \ldots, (1, t_m, \ldots, t_m^m) \in \mathbb{R}^{m+1}$ are linearly independent. But this is indeed the case , since the Vandermonde determinant

$$\det \begin{pmatrix} 1 & t_0 & t_0^2 & \cdots & t_0^m \\ 1 & t_1 & t_0^2 & \cdots & t_1^m \\ & & \vdots & & \\ 1 & t_m & t_m^2 & \cdots & t_m^m \end{pmatrix} = \prod_{0 \leq i < j \leq m} (t_j - t_i) \neq 0$$

is nonzero.   □

Proposition B.41. *Any abstract simplicial complex* K *of dimension* d *admits a geometric realization in* $\mathbb{R}^{2d+1}$.

Proof. Let $\text{vert}(K) = \{v_1, \ldots, v_r\}$ be the vertex set of K and $\gamma : \mathbb{R} \to \mathbb{R}^{2d+1}$ the moment curve in $\mathbb{R}^{2d+1}$. Define the geometric simplicial complex Δ with vertices $\{\gamma(1), \ldots, \gamma(r)\}$ by

$$\Delta = \{\text{conv}(\gamma(i_0), \ldots, \gamma(i_k)) : \{v_{i_0}, \ldots, v_{i_k}\} \in K\}.$$

We have to show that Δ satisfies the requirements of a simplicial complex. Clearly it is closed under taking faces. Now the union $\{v_{i_0}, \ldots, v_{i_k}\} \cup \{v_{j_0}, \ldots, v_{j_l}\}$ of the vertex sets of two simplices in K has cardinality at most $2d + 2$. By the previous lemma, the corresponding set $\{\gamma(i_0), \ldots, \gamma(i_k)\} \cup \{\gamma(j_0), \ldots, \gamma(j_l)\}$ is affinely independent, and hence, by Lemma B.40, the two simplices $\mathrm{conv}(\gamma(i_0), \ldots, \gamma(i_k))$ and $\mathrm{conv}(\gamma(j_0), \ldots, \gamma(j_l))$ intersect in the expected common face.

Finally, the map $\varphi : \mathrm{vert}(K) \longrightarrow \mathrm{vert}(\Delta)$ defined by $v_i \longmapsto \gamma(i)$ defines a bijection of the vertex sets that induces the necessary correspondence of simplices.

□

The following proposition justifies concentrating on abstract simplicial complexes for all combinatorial purposes. In particular, it implies that a geometric realization of an abstract simplicial complex is unique up to simplicial homeomorphisms.

Proposition B.42. *Any two geometric simplicial complexes are simplicially homeomorphic if and only if the associated abstract simplicial complexes are simplicially isomorphic.*   □

For convenience, we will frequently make use of the notation $|K|$ for some geometric realization of the abstract simplicial complex K and $|\sigma|$ for the geometric simplex corresponding to $\sigma \in K$. Moreover, the simplicial map of geometric complexes induced by a simplicial map $f : K \to L$ will be denoted by $|f| : |K| \to |L|$. This slight abuse of notation is justified by the previous proposition.

B.4 Shellability of Simplicial Complexes

A convenient way to prove that the polyhedron of a d-dimensional simplicial complex is $(d - 1)$-connected is to show that it is *shellable* [Bjö94]. There are many different concepts of shellability. Here we introduce a form of shellability that will be referred to as *topological shelling*.

Definition B.43. Suppose that Δ is a simplicial complex. A total ordering $\sigma_1, \sigma_2, \ldots, \sigma_k$ of its (inclusion) maximal faces is a *topological shelling* of Δ if

$$\dim(\sigma_1) \geq \dim(\sigma_2) \geq \cdots \geq \dim(\sigma_k)$$

and for each $j > 1$, either (a) or (b) is satisfied, where

1. $(\bigcup_{i<j} \sigma_i) \cap \sigma_j$ is a contractible subset of $\partial\sigma_j$;
2. $(\bigcup_{i<j} \sigma_i) \cap \sigma_j = \partial\sigma_j$.

More often than not, conditions (a) and (b) are combined into a single stronger condition, namely into

(*) $(\bigcup_{i<j} \sigma_i) \cap \sigma_j$ is a pure $(\dim(\sigma_j) - 1)$-dimensional complex.

The homotopy type of a simplicial complex possessing a topological shelling is now easy to determine.

Proposition B.44. *Suppose that Δ is a simplicial complex that admits a topological shelling $\sigma_1, \sigma_2, \ldots, \sigma_k$, $k \geq 1$. Let $n_i = \dim(\sigma_i)$. Then $|\Delta|$ is homotopic to the wedge of spheres, $\bigvee_{j \in J} \mathbb{S}^{n_j}$, where $J = \{j : (\bigcup_{i<j} \sigma_i) \cap \sigma_j = \partial \sigma_j\}$.*

Proof (Sketch). We proceed by induction on the number k of maximal simplices of Δ. The case $k = 1$ is clear. For the induction step with $k \geq 2$, let P and $P_{<k}$ be the polyhedra $P = |\Delta| = \bigcup_{i \leq k} \sigma_i$ and $P_{<k} = \bigcup_{i < k} \sigma_i$, respectively.

By the induction hypothesis, $P_{<k}$ is homotopic to a wedge of spheres $\bigvee_{j \in J, j < k} \mathbb{S}^{n_j}$.

Now if, as in condition (a), $P_{<k} \cap \sigma_k$ is a contractible subset of $\partial \sigma_k$, then $P_{<k}$ and P have the same homotopy type and $k \notin J$.

If, as in condition (b), $P_{<k} \cap \sigma_k = \partial \sigma_k$, then note that

$$n_1 \geq n_2 \geq \cdots \geq n_{k-1} \geq n_k = \dim(\sigma_k) > \dim(\partial \sigma_k)$$

implies that any image of $\partial \sigma_k$ is contractible in $P_{<k} \simeq \bigvee_{j \in J, j < k} \mathbb{S}^{n_j}$, and hence $P \simeq \bigvee_{j \in J} \mathbb{S}^{n_j}$. $\qquad \square$

Recall that a simplicial complex is called *pure n-dimensional* if all maximal faces are of the same dimension n.

Corollary B.45. *A pure n-dimensional simplicial complex Δ admitting a topological shelling is $(n-1)$-connected.* $\qquad \square$

B.5 Some Operations on Simplicial Complexes

Barycentric Subdivision

One of the most important operations that can be performed on a simplicial complex is that of *barycentric subdivision*, i.e., a simultaneous "refinement" of all simplices.

The operation is easily described for an abstract simplicial complex K. The *(first) barycentric subdivision of K* is defined to be the complex

$$\operatorname{sd} K = \{\{\sigma_0, \ldots, \sigma_k\} : \sigma_0, \ldots, \sigma_k \in K \setminus \{\emptyset\}, \sigma_0 \subset \sigma_1 \subset \cdots \subset \sigma_k\}.$$

In other words, the simplices of K become the vertices of $\operatorname{sd} K$—we think of the barycenters of the simplices in K—and any inclusion chain of simplices in K defines a simplex of $\operatorname{sd} K$. Note that the empty inclusion chain in the case $k = -1$ yields the empty simplex in $\operatorname{sd} K$. The construction is easily understood by looking at Fig. B.6. The first example shows the barycentric subdivision of the complex K given by a 1-dimensional simplex. In this case we have

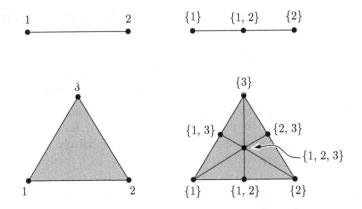

Fig. B.6 Examples of the barycentric subdivision of a 1- and a 2-dimensional simplex

$$K = \{\emptyset, \{1\}, \{2\}, \{1, 2\}\},$$

$$\text{sd } K = \{\emptyset, \{\{1\}\}, \{\{2\}\}, \{\{1, 2\}\}, \{\{1\}, \{1, 2\}\}, \{\{2\}, \{1, 2\}\}\}.$$

This procedure can certainly be iterated, and we will denote the *kth barycentric subdivision of K* by $\text{sd}^k K$.

Now we will briefly discuss the geometric situation. If $\sigma = \text{conv}(x_0, \ldots, x_n) \subseteq \mathbb{R}^m$ is a geometric simplex, then its *barycenter* $b(\sigma)$ is defined by

$$b(\sigma) = \frac{1}{n+1} \sum_{i=0}^{n} x_i.$$

Now, if Δ is a geometric simplicial complex, then its barycentric subdivision is given by the complex

$$\text{sd } \Delta = \{\text{conv}(b(\sigma_0), \ldots, b(\sigma_k)) : \sigma_0, \ldots, \sigma_k \in \Delta \setminus \{\emptyset\}, \sigma_0 \subset \cdots \subset \sigma_k\}.$$

Clearly, if K is the associated abstract simplicial complex of Δ, then the associated abstract simplicial complex of $\text{sd } \Delta$ is simplicially isomorphic to $\text{sd } K$.

Proposition B.46. *If Δ is a geometric simplicial complex, then the simplicial complex $\text{sd } \Delta$ is a subdivision of Δ according to Definition B.36.*   □

Proposition B.47. *Let $\sigma_0 \subset \sigma_1 \subset \cdots \subset \sigma_n$ be an inclusion chain of geometric simplices, where $\dim(\sigma_i) = i$, for $i = 0, \ldots, n$. Then we have*

$$\text{diam}(\text{conv}(b(\sigma_0), \ldots, b(\sigma_n))) \le \frac{n}{n+1} \text{diam}(\sigma_n).$$ □

Corollary B.48. *If Δ is a geometric simplicial complex of dimension n, then*

$$\max_{\tau \in \mathrm{sd}\, \Delta} \mathrm{diam}(\tau) \leq \frac{n}{n+1} \max_{\sigma \in \Delta} \mathrm{diam}(\sigma).$$

□

Joins

Definition B.49. Let K and L be abstract simplicial complexes. Define their *join,* $K * L$, by

$$K * L = \{\sigma + \tau : \sigma \in K, \tau \in L\},$$

where the *join of simplices* $\sigma + \tau$, also denoted by $\sigma * \tau$, denotes the set sum as defined on page 166. If σ is a simplex, then the *join* $\sigma * K$ *of* σ *with* K is defined by

$$\sigma * K = \{\tau + \rho : \tau \leq \sigma, \rho \in K\}.$$

Note that the associated abstract simplicial complex of the boundary complex Γ^n of the cross polytope may be identified with the $(n+1)$-fold iterated join of the complex given by two isolated vertices.

A join can just as well be defined for geometric simplicial complexes Δ and Γ such that it is compatible with the associated abstract structure, i.e., $K(\Delta * \Gamma) \cong K(\Delta) * K(\Gamma)$.

B.6   The Language of Category Theory

Category theory provides the proper language to introduce and efficiently handle algebraic invariants of topological spaces.

Categories

A *category* $\mathcal{C}$ consists of

- A collection of *objects*;
- For each pair A, B of objects, a set of *morphisms,* $\mathcal{C}(A, B)$, between them;
- For each object A, an *identity morphism* $\mathrm{id}_A \in \mathcal{C}(A, A)$;
- A *composition law*

$$\circ : \mathcal{C}(B, C) \times \mathcal{C}(A, B) \longrightarrow \mathcal{C}(A, C)$$

for each triple A, B, C of objects.

The composition law is required to be associative, i.e.,

$$h \circ (g \circ f) = (h \circ g) \circ f,$$

and the identity has to satisfy the expected equalities,

$$\mathrm{id} \circ f = f \text{ and } f \circ \mathrm{id} = f,$$

whenever these compositions are defined. A category is *small* if its collection of objects is a set. Typically, a morphism $f \in C(A, B)$ is denoted by $f : A \longrightarrow B$.

Some examples of categories are the following: the category S with sets as objects and set maps as morphisms; the category $\mathcal{TOP}$ of topological spaces and continuous maps; the category $\mathcal{G}$ of groups and homomorphisms; the category $\mathcal{A}$ of abelian groups and homomorphisms; and the category $\mathcal{V}$ of F-vector spaces and F-linear maps for some fixed field F. In all these cases, the composition law is given by the usual composition of maps, and the identity morphisms are given by the actual set identity maps.

Functors

A *covariant functor* $F : C \longrightarrow D$ of categories is an assignment of objects and morphisms from C to D. More precisely, each object A of C is assigned an object $F(A)$ of D, and each morphism $f : A \longrightarrow B$ of C is assigned to a morphism $F(f) : F(A) \longrightarrow F(B)$ of D such that

$$F(\mathrm{id}_A) = \mathrm{id}_{F(A)} \text{ and } F(g \circ f) = F(g) \circ F(f).$$

A simple example is the *forgetful functor* $F : \mathcal{TOP} \longrightarrow S$ from the category of topological spaces to the category of sets, which assigns to each topological space $(X, \mathcal{O})$ the underlying set X, and to each continuous map $f : (X, \mathcal{O}_X) \longrightarrow (Y, \mathcal{O}_Y)$ the set map $f : X \longrightarrow Y$.

There is also the notion of a *contravariant functor* $F : C \longrightarrow D$, which behaves similarly to a covariant functor, but turns arrows around, i.e., it is given by assigning an object $F(A)$ from D to each object $A \in C$, and by an assigment $F(f) : F(B) \longrightarrow F(A)$ to each morphism $f \in C(A, B)$ such that

$$F(\mathrm{id}_A) = \mathrm{id}_{F(A)} \text{ and } F(g \circ f) = F(f) \circ F(g).$$

An interesting example, well known from linear algebra, is the functor $* : \mathcal{V} \longrightarrow \mathcal{V}$ assigning to each F-vector space V its dual space $V^* = \{\varphi : V \longrightarrow F : \varphi \text{ linear}\}$ and to each linear map $f : V \longrightarrow W$ its dual map $f^* : W^* \longrightarrow V^*$ defined by $\varphi \mapsto \varphi \circ f$.

Natural Transformations

A *natural transformation* α between (covariant) functors $F, G : \mathcal{C} \longrightarrow \mathcal{D}$ consists of morphisms, $\alpha_A : F(A) \longrightarrow G(A)$, for each object A of $\mathcal{C}$ such that for each morphism $f : A \longrightarrow B$ in $\mathcal{C}$, the following diagram commutes:

$$
\begin{array}{ccc}
F(A) & \xrightarrow{\ F(f)\ } & F(B) \\
\alpha_A \downarrow & \cdot & \downarrow \alpha_B \\
G(A) & \xrightarrow{\ G(f)\ } & G(B)
\end{array}
$$

For an example we return to the dualization functor $* : \mathcal{V} \longrightarrow \mathcal{V}$ from linear algebra. Recall that a finite-dimensional vector space V is isomorphic to its dual V^*. But such an isomorphism depends on a choice of a basis: it is not *natural*. In contrast, there is a *natural* way to define an isomorphism between V and its second dual V^{**} by defining $v \mapsto i(v)$, where $i_v(\varphi) = \varphi(v)$. In fact, i defines a natural transformation between the two covariant functors id and $**$.

Two categories $\mathcal{C}$ and $\mathcal{D}$ are defined to be *equivalent* if there are functors $F : \mathcal{C} \longrightarrow \mathcal{D}$ and $G : \mathcal{D} \longrightarrow \mathcal{C}$ and natural transformations $\alpha : G \circ F \longrightarrow \mathrm{id}_{\mathcal{C}}$ and $\beta : F \circ G \longrightarrow \mathrm{id}_{\mathcal{D}}$ to the respective identity functors of $\mathcal{C}$ and $\mathcal{D}$.

B.7 Some Homological Algebra

This section provides some algebraic concepts that are needed for the introduction of algebraic invariants for topological spaces.

Chain Complexes

Definition B.50. A *graded (abelian) group* is a collection of abelian groups C_i, $i \in \mathbb{Z}$.

Definition B.51. A *chain complex* is a graded group $\{C_i : i \in \mathbb{Z}\}$ together with a sequence of homomorphisms $\partial_i : C_i \longrightarrow C_{i-1}$ such that $\partial_{i-1} \circ \partial_i : C_i \longrightarrow C_{i-2}$ is the zero homomorphism. Each ∂_i is called a *boundary operator*.

Abusing notation, we will abbreviate ∂_i by ∂ and write ∂^2 for $\partial_{i-1} \circ \partial_i$, and so on. For a chain complex, we will abbreviate the pair $C_* = (\{C_i : i \in \mathbb{Z}\}, \partial)$.

Definition B.52. Let C_* and D_* be chain complexes. A *chain map* f from C_* to D_* is a collection $\{f_i : i \in \mathbb{Z}\}$ of homomorphisms $f_i : C_i \longrightarrow D_i$ such that $\partial \circ f_i = f_{i-1} \circ \partial$, i.e., the diagram

$$C_i \xrightarrow{\ \partial\ } C_{i-1}$$

$$f_i \downarrow \qquad\qquad \downarrow f_{i-1}$$

$$D_i \xrightarrow{\ \partial\ } D_{i-1}$$

commutes for each i.

Definition B.53. If C_* is a chain complex, then we define its *homology* to be the graded group $\{H_i : i \in \mathbb{Z}\}$ given by the quotients

$$H_i(C_*) = \frac{\ker(\partial_i : C_i \to C_{i-1})}{\mathrm{im}\,(\partial_{i+1} : C_{i+1} \to C_i)}.$$

The elements of $Z_i = \ker(\partial_i : C_i \to C_{i-1})$ are called the *cycles*, while the elements of $B_i = \mathrm{im}\,(\partial_{i+1} : C_{i+1} \to C_i)$ are called the *boundaries* in C_i. Each element $[z] = z + B_i$ of the homology group $H_p(C_*)$ is represented by a cycle $z \in Z_i$.

Proposition B.54. *A chain map* $f : C_* \longrightarrow D_*$ *induces a map,* f_*, *in homology, i.e., homomorphisms* $f_* : H_p(C_*) \longrightarrow H_p(D_*)$ *for each* p *that are defined by* $f_*([z]) = [f(z)]$. *Clearly, if* id $: C_* \longrightarrow C_*$ *is the identity chain map, then* $\mathrm{id}_* = \mathrm{id}_{H_*(C_*)}$, *and if* $g : D_* \longrightarrow E_*$ *is another chain map, then* $(g \circ f)_* = g_* \circ f_*$. $\square$

In other words, we have just seen that *homology* is a functor from the category of chain complexes with chain maps to the category of graded groups with grade-preserving homomorphisms.

Exact Sequences

Definition B.55. A sequence

$$A \xrightarrow{i} B \xrightarrow{j} C$$

of homomorphisms is *exact* if $\mathrm{im}\,i = \ker j$.

With this terminology one could say that homology is a measure of exactness. In particular, $H_p(C_*) = 0$ if and only if the sequence

$$C_{p+1} \xrightarrow{\partial} C_p \xrightarrow{\partial} C_{p-1}$$

is exact.

A sequence

$$A_0 \xrightarrow{i_0} A_1 \xrightarrow{i_1} A_2 \xrightarrow{i_2} \cdots \xrightarrow{i_n} A_{n+1}$$

is called exact if each subsequence

$$A_{k-1} \xrightarrow{i_{k-1}} A_k \xrightarrow{i_k} A_{k+1}$$

is exact, $k = 1, \ldots, n$.

A *short exact sequence* is an exact sequence of the type

$$0 \longrightarrow A \xrightarrow{i} B \xrightarrow{j} C \longrightarrow 0.$$

Theorem B.56. *If* $A_* \xrightarrow{i} B_* \xrightarrow{j} C_*$ *is a sequence of chain maps such that*

$$0 \longrightarrow A_k \xrightarrow{i} B_k \xrightarrow{j} C_k \longrightarrow 0$$

is a short exact sequence for each k, *then there exist* connecting homomorphisms $\partial_p : H_p(C_*) \longrightarrow H_{p-1}(A_*)$ *inducing a* long exact sequence in homology:

$$\cdots \xrightarrow{j_*} H_{p+1}(C_*) \xrightarrow{\partial_{p+1}} H_p(A_*) \xrightarrow{i_*} H_p(B_*) \xrightarrow{j_*} H_p(C_*) \xrightarrow{\partial_p} H_{p-1}(A_*) \xrightarrow{i_*} \cdots .$$

Proof. We will only provide the construction of the connecting homomorphisms. The proof of the exactness of the long sequence is the content of an exercise.

Let $[z] \in H_p(C_*)$, $z \in \ker(\partial_p : C_p \to C_{p-1})$. Since j_p is surjective, there exists an element $b \in B_p$ with $j_p(b) = z$. The fact that $j_{p-1}\partial_p(b) = \partial_p j_p(b) = \partial_p(z) = 0$ shows that $\partial_p(b) \in \ker j_{p-1} = \operatorname{im} i_{p-1}$. Hence, by injectivity of i_{p-1}, there exists a unique element $a \in A_{p-1}$ such that $i_{p-1}(a) = \partial_p(b)$. The reader is urged to take a pencil and to make notes in the following diagram:

$$
\begin{array}{ccccccc}
 & & B_{p+1} & \xrightarrow{j_{p+1}} & C_{p+1} & \longrightarrow & 0 \\
 & & \downarrow{\scriptstyle \partial_{p+1}} & & \downarrow{\scriptstyle \partial_{p+1}} & & \\
0 \longrightarrow & A_p & \xrightarrow{i_p} & B_p & \xrightarrow{j_p} & C_p & \longrightarrow 0 \\
 & \downarrow{\scriptstyle \partial_p} & & \downarrow{\scriptstyle \partial_p} & & \downarrow{\scriptstyle \partial_p} & \\
0 \longrightarrow & A_{p-1} & \xrightarrow{i_{p-1}} & B_{p-1} & \xrightarrow{j_{p-1}} & C_{p-1} & \longrightarrow 0 \\
 & \downarrow{\scriptstyle \partial_{p-1}} & & \downarrow{\scriptstyle \partial_{p-1}} & & & \\
0 \longrightarrow & A_{p-2} & \xrightarrow{i_{p-2}} & B_{p-2} & & &
\end{array}
$$

Now define $\partial_p : H_p(C_*) \longrightarrow H_{p-1}(A_*)$ by $\partial_p([z]) = [a]$. We have to show that this is well defined, i.e., that a is a cycle in A_{p-1}, and that the definition of $[a]$ is independent of

- The choice b of a preimage of z under j_p, and
- The choice of a representative within the class $[z]$.

The computation $i_{p-2}\partial_{p-1}(a) = \partial_{p-1}i_{p-1}(a) = \partial_{p-1}\partial_p(b) = 0$ and the injectivity of i_{p-2} show that a is a cycle.

Now assume that $b, b' \in B_p$ are such that $j_p(b) = j_p(b') = z$. Then $b' - b \in \ker j_p = \operatorname{im} i_p$, and hence there exists an element $\bar{a}$ such that $i_p(\bar{a}) = b' - b$. Now $i_{p-1}\partial_p(\bar{a}) = \partial_p i_p(\bar{a}) = \partial_p(b' - b) = \partial_p(b') - \partial_p(b)$. Hence $i_{p-1}^{-1}(\partial_p(b')) - i_{p-1}^{-1}(\partial_p(b)) = \partial_p(\bar{a})$ is a boundary in A_{p-1}.

Finally, let $[z] = [z']$ and assume $j_p(b) = z$. Since $z' - z$ is a boundary in C_p, there exists a $c \in C_{p+1}$ such that $\partial_{p+1}(c) = z' - z$. By surjectivity of j_{p+1}, there exists an element $\bar{b} \in B_{p+1}$ with $j_{p+1}(\bar{b}) = c$. Set $b' = b + \partial_{p+1}(\bar{b}) \in B_p$. Then

$$j_p(b') = j_p(b) + j_p\partial_{p+1}(\bar{b}) = z + \partial_{p+1}j_{p+1}(\bar{b}) = z + z' - z = z',$$

making b' a bona fide choice as a preimage of z'. But then $\partial_p(b') = \partial_p(b)$, since $\partial^2 = 0$. Hence, $\partial_p(b')$ and $\partial_p(b)$ also have the same preimage under the map i_{p-1}.
$\square$

B.8 Axioms for Homology

Homology theory is a quite general concept that features the essential idea of algebraic topology: assigning algebraic invariants to topological spaces. It appears in many guises in various contexts in mathematics. Here we will state the axioms for homology, and in the next section we will discuss the particular example of simplicial homology.

Let $\mathcal{A}$ be the category of abelian groups and let $\mathcal{X}$ be a category of pairs of topological spaces, e.g., pairs of polyhedra together with continuous maps.

Definition B.57. A *homology theory on* $\mathcal{X}$ is given by

- A family $\{H_p : p \in \mathbb{Z}\}$ of functors from $\mathcal{X}$ to $\mathcal{A}$ assigning to each integer p and each pair (X, A) of topological spaces (of $\mathcal{X}$) an abelian group $H_p(X, A)$, and to each $f : (X, A) \longrightarrow (Y, B)$ a group homomorphism $f_* = H_p(f) : H_p(X, A) \longrightarrow H_p(Y, B)$;
- A family $\{\partial_p : p \in \mathbb{Z}\}$ of natural transformations $\partial_p : H_p(X, A) \longrightarrow H_{p-1}(A)$, where $H_{p-1}(A)$ abbreviates $H_{p-1}(A, \emptyset)$,

such that the following axioms are satisfied:

(Dimension) If P is the one-point space, $P = \{*\}$, then $H_q(P) = 0$ for all $q \neq 0$.
(Homotopy) If $f \simeq g : (X, A) \longrightarrow (Y, B)$ are homotopic, then

$$f_* = g_* : H_p(X, A) \longrightarrow H_p(Y, B).$$

(Exactness) If for a pair (X, A), the obvious inclusions are denoted by $i : (A, \emptyset) \hookrightarrow (X, \emptyset)$ and $j : (X, \emptyset) \hookrightarrow (X, A)$, then the following long sequence is exact:

$$\cdots \xrightarrow{j_*} H_{p+1}(X, A) \xrightarrow{\partial_{p+1}} H_p(A) \xrightarrow{i_*} H_p(X) \xrightarrow{j_*} H_p(X, A) \xrightarrow{\partial_p} H_{p-1}(A) \xrightarrow{i_*} \cdots .$$

(Excision) If (X, A) is a pair, and $U \subseteq X$ is an open set such that $\mathrm{cl}(U) \subseteq \mathrm{int}(A)$, then the inclusion $k : (X \setminus U, A \setminus U) \hookrightarrow (X, A)$ induces an isomorphism

$$k_* : H_p(X \setminus U, A \setminus U) \xrightarrow{\cong} H_p(X, A).$$

(Additivity) If $(X, A) = (\coprod_\alpha X_\alpha, \coprod_\alpha A_\alpha)$ is a direct sum of pairs (X_α, A_α), and $i_\alpha : (X_\alpha, A_\alpha) \hookrightarrow (X, A)$ are the inclusions, then the sum of homomorphisms

$$\oplus_\alpha (i_\alpha)_* : \bigoplus_\alpha H_p(X_\alpha, A_\alpha) \xrightarrow{\cong} H_p(X, A)$$

is an isomorphism.

Definition B.58. For any homology theory, the group $H_0(P) = G$, where P is the one-point space, is called the *coefficient group of the theory.*

Many properties can be derived from the axioms just by algebraic reasoning.

Proposition B.59. *If $f : (X, A) \longrightarrow (Y, B)$ is a homotopy equivalence of pairs, then it induces an isomorphism*

$$f_* : H_p(X, A) \xrightarrow{\cong} H_p(Y, B).$$

Proof. Let $g : (Y, B) \longrightarrow (X, A)$ be a homotopy inverse to f. By definition, $g \circ f \simeq \mathrm{id}_{(X,A)}$ and $f \circ g \simeq \mathrm{id}_{(Y,B)}$, and hence $g_* \circ f_* = (g \circ f)_* = (\mathrm{id}_{(X,A)})_* = \mathrm{id}_{H_p(X,A)}$ by the functorial properties of the homology functor and the homotopy axiom. Analogously, $f_* \circ g_* = \mathrm{id}_{H_p(Y,B)}$.   □

Corollary B.60. *If X is a contractible space, then the homology of X coincides with the homology of the one-point space P.*   □

This class of spaces leads us to an important notion.

Definition B.61. If the coefficient group of the homology theory is G, i.e., $H_0(P) = G$, then a space X is called G-*acyclic* if $H_*(X) = H_*(P)$.

Reduced Homology

For most computations it is convenient to consider *reduced homology*, which is easily obtained from any homology theory.

Let X be a nonempty space and $\epsilon : X \longrightarrow P$ the unique map to the one-point space P. The homomorphism $\epsilon_* : H_0(X) \longrightarrow H_0(P) = G$ induced by ϵ is called the *augmentation map*. Since for any $i : P \hookrightarrow X$, the composition $\epsilon \circ i$ is the identity map id_P, we have $\epsilon_* \circ i_* = \mathrm{id}_{H_0(P)}$, and therefore ϵ_* is surjective. By definition, the sequence

$$0 \longrightarrow \ker(\epsilon_*) \longrightarrow H_0(X) \xrightarrow{\epsilon_*} H_0(P) \longrightarrow 0$$

is short exact, and there is an isomorphism $H_0(X) \cong \ker(\epsilon_*) \oplus G$ (which is not natural in X, though).

Now set $\widetilde{H}_0(X) = \ker(\epsilon_*)$ and $\widetilde{H}_p(X) = H_p(X)$ for $p \neq 0$. Furthermore, for $A \neq \emptyset$, we set $\widetilde{H}_p(X, A) = H_p(X, A)$. Note that with this new terminology, $\widetilde{H}_*(P) = 0$. In particular, a space X is G-acyclic if $\widetilde{H}_*(X) = 0$.

Proposition B.62. *If (X, A) is such that $A \neq \emptyset$, then the following long sequence is exact:*

$$\cdots \xrightarrow{j_*} \widetilde{H}_{p+1}(X, A) \xrightarrow{\partial_{p+1}} \widetilde{H}_p(A) \xrightarrow{i_*} \widetilde{H}_p(X) \xrightarrow{j_*} \widetilde{H}_p(X, A) \xrightarrow{\partial_p} \widetilde{H}_{p-1}(A) \xrightarrow{i_*} \cdots . \quad \square$$

It is not very hard to compute the homology groups of spheres in general.

Theorem B.63. *Let G be the coefficient group of the homology theory. Then the homology of the n-dimensional sphere $\mathbb{S}^n$ is given by*

$$\widetilde{H}_p(\mathbb{S}^n) = \begin{cases} G, & \text{if } p = n, \\ 0, & \text{if } p \neq n. \end{cases} \qquad \square$$

B.9 Simplicial Homology

We will define a homology theory for the category of pairs of polyhedra together with continuous maps. In order to do so, we will first concentrate on pairs of simplicial complexes and simplicial maps.

The Oriented Simplicial Chain Complex

Let K be an (abstract) simplicial complex. Consider an ordering $\mathrm{vert}(K) = \{v_0, v_1, \ldots\}$ of the vertices of K. For $n \geq 0$, define the nth *oriented simplicial chain group (with integer coefficients)*, $C_n(K)$, to be the free abelian group generated by the set of *oriented n-simplices*

$$\{\langle v_{i_0}, \ldots, v_{i_n} \rangle : \{v_{i_0}, \ldots, v_{i_n}\} \in K, \dim(\{v_{i_0}, \ldots, v_{i_n}\}) = n\}$$

modulo the relations

$$\langle v_{i_0}, \ldots, v_{i_n} \rangle = \text{sign}(\pi)\langle v_{i_{\pi(0)}}, \ldots, v_{i_{\pi(n)}} \rangle,$$

where $\pi : \{0, \ldots, n\} \longrightarrow \{0, \ldots, n\}$ is an arbitrary permutation. For $n < 0$, set $C_n(K) = 0$. We think of an element of $C_n(K)$ as a set of oriented n-simplices of K each labeled with an integer.

For later use we note that in the baby example given by the one-point simplicial complex $P = \{\emptyset, \{p\}\}$, the chain groups compute to

$$C_n(P) = \begin{cases} \mathbb{Z} \cdot \langle p \rangle, & \text{if } n = 0, \\ 0, & \text{if } n \neq 0. \end{cases}$$

We will now make the graded abelian group $\{C_i(K) : i \in \mathbb{Z}\}$ into a chain complex by defining boundary operators $\partial : C_n(K) \longrightarrow C_{n-1}(K)$ by

$$\partial(\langle v_{i_0}, \ldots, v_{i_n} \rangle) = \sum_{k=0}^{n} (-1)^k \langle v_{i_0}, \ldots, \hat{v}_{i_k}, \ldots, v_{i_n} \rangle,$$

where $\hat{v}_{i_k}$ denotes the omission of the vertex v_{i_k}, i.e.,

$$\langle v_{i_0}, \ldots, \hat{v}_{i_k}, \ldots, v_{i_n} \rangle = \langle v_{i_0}, \ldots, v_{i_{k-1}}, v_{i_{k+1}}, \ldots, v_{i_n} \rangle.$$

Proposition B.64. $\partial : C_n(K) \longrightarrow C_{n-1}(K)$ *is a well-defined homomorphism with* $\partial^2 = 0$. □

The pair $C_*(K) = (\{C_n(K) : n \in \mathbb{Z}\}, \partial)$ is the *oriented simplicial chain complex* of K.

The Oriented Simplicial Chain Complex with Field Coefficients

We just defined the oriented simplicial chain complex with integer coefficients. One can define any abelian group G to be the coefficient group of our chain complex by considering the chain groups $C_n(K) \otimes G$ together with the boundary operator $\partial \otimes \text{id}$. For the purpose of this book, however, we will need only the additional case of field coefficients, which is slightly easier to describe.

Let F be a field. For $n \geq 0$, we define the nth *oriented simplicial chain group* $C_n(K; F)$ *with coefficients in the field* F to be the F-vector space generated by the set of *oriented n-simplices*

$$\{\langle v_{i_0}, \ldots, v_{i_n} \rangle : \{v_{i_0}, \ldots, v_{i_n}\} \in K, \dim(\{v_{i_0}, \ldots, v_{i_n}\}) = n\}$$

modulo the subspace generated by the relations

$$\langle v_{i_0}, \dots, v_{i_n} \rangle = \text{sign}(\pi) \langle v_{i_{\pi(0)}}, \dots, v_{i_{\pi(n)}} \rangle,$$

where $\pi : \{0, \dots, n\} \longrightarrow \{0, \dots, n\}$ is an arbitrary permutation. For $n < 0$, set $C_n(K; F) = 0$. The boundary maps are defined just as in the integer coefficients case. We denote the resulting chain complex by $C_*(K; F) = (\{C_n(K; F) : n \in \mathbb{Z}\}, \partial)$ and the resulting homology groups $\{H_n(C_*(K; F)) : n \in \mathbb{Z}\}$ by $H_*(K; F)$.

The Oriented Simplicial Chain Complex for Pairs

Consider a pair (K, L) of simplicial complexes. Clearly the group $C_n(L)$ constitutes a subgroup of $C_n(K)$, and hence we may define the *oriented simplicial chain groups of the pair* (K, L) by

$$C_n(K, L) = C_n(K)/C_n(L).$$

The boundary operator of $C_n(K)$ defines a boundary operator $\partial : C_n(K, L) \longrightarrow C_{n-1}(K, L)$, and we obtain the *oriented simplicial chain complex* $C_*(K, L) = (\{C_n(K, L) : n \in \mathbb{Z}\}, \partial)$.

Let $i : C_*(L) \hookrightarrow C_*(K)$ be the chain map defined by the inclusions $C_n(L) \hookrightarrow C_n(K)$, and let $\pi : C_*(K) \longrightarrow C_*(K, L)$ be the quotient chain map. Then the following short sequence is exact:

$$0 \longrightarrow C_*(L) \overset{i}{\longrightarrow} C_*(K) \overset{\pi}{\longrightarrow} C_*(K, L) \longrightarrow 0.$$

Therefore, by Theorem B.56, there exists a family $\{\partial_p : p \in \mathbb{Z}\}$ of connecting homomorphisms inducing the long exact homology sequence

$$\cdots \overset{\pi_*}{\longrightarrow} H_{p+1}(K, L) \overset{\partial_{p+1}}{\longrightarrow} H_p(L) \overset{i_*}{\longrightarrow} H_p(K) \overset{\pi_*}{\longrightarrow} H_p(K, L) \overset{\partial_p}{\longrightarrow} H_{p-1}(L) \overset{i_*}{\longrightarrow} \cdots,$$

where we used the abbreviations $H_p(K) = H_p(C_*(K))$ and $H_p(K, L) = H_p(C_*(K, L))$ and so on.

We have now encountered a first link between simplicial complexes and associated homology groups. This link satisfies the following property with respect to barycentric subdivision.

Theorem B.65. *Let* (K, L) *be a pair of simplicial complexes. Then there exists a (natural) chain map* $C_*(K, L) \longrightarrow C_*(\text{sd } K, \text{sd } L)$ *that induces isomorphisms* $H_p(K, L) \overset{\cong}{\longrightarrow} H_p(\text{sd } K, \text{sd } L)$ *in all dimensions.* □

Unfortunately, the proof of this theorem would lead us too far beyond the scope of the book.

As we saw before, we may use the same definitions for field coefficients and obtain the chain complex $C_*(K, L; F)$, the homology groups $H_p(K, L; F)$, the corresponding long exact sequence, and so on.

From Simplicial to Chain Maps

Let K and L be simplicial complexes and $f : K \longrightarrow L$ a simplicial map. We define the induced chain map $f_\Delta : C_*(K) \longrightarrow C_*(L)$ by

$$f_\Delta : C_n(K) \longrightarrow C_n(L),$$

$$\langle v_{i_0}, \ldots, v_{i_n} \rangle \longmapsto \begin{cases} \langle f(v_{i_0}), \ldots, f(v_{i_n}) \rangle, & \text{if } \dim(\{ f(v_{i_0}), \ldots, f(v_{i_n}) \}) = n, \\ 0, & \text{otherwise.} \end{cases}$$

It is an easy exercise to check that this is indeed a chain map and, as such, induces a map f_* in homology $f_* : H_n(K) \longrightarrow H_n(L)$. Now it is not hard to show that the connecting homomorphisms behave naturally with respect to such induced maps.

Proposition B.66. *If $f : (K, L) \longrightarrow (K', L')$ is a simplicial map, then the following diagram commutes:*

$$
\begin{array}{ccc}
H_p(K, L) & \xrightarrow{\ \partial_p\ } & H_{p-1}(L) \\
{\scriptstyle f_*}\downarrow & & \downarrow{\scriptstyle f_*} \\
H_p(K', L') & \xrightarrow{\ \partial_p\ } & H_{p-1}(L')
\end{array}
$$

$\square$

Reduced Simplicial Homology

In the case of simplicial homology, we may define an *augmentation map* on the chain complex level inducing the augmentation map in homology. If K is a simplicial complex, then define $\epsilon : C_0(K) \longrightarrow \mathbb{Z}$ by $\epsilon(\sum_{k=1}^{m} n_k v_{i_k}) = \sum_{k=1}^{m} n_k$. Then we can define the reduced homology groups $\tilde{H}_*$ as the homology groups of the chain complex

$$\cdots \xrightarrow{\partial} C_n(K) \xrightarrow{\partial} \cdots \xrightarrow{\partial} C_1 \xrightarrow{\partial} C_0(K) \xrightarrow{\epsilon} \mathbb{Z} \longrightarrow 0.$$

Note that indeed, the augmentation map on the chain complex level is essentially induced by the simplicial map $K \longrightarrow P$ to the one-point complex $P = \{\emptyset, \{p\}\}$.

From Continuous to Simplicial Maps

In order to define a homology theory, we need to be able to deal with continuous maps of polyhedra. Let Δ and Γ be geometric simplicial complexes.

Definition B.67. For $x \in |\Delta|$, define the *carrier,* carr(x), to be the inclusionwise smallest simplex of Δ containing x.

Definition B.68. If $f : |\Delta| \longrightarrow |\Gamma|$ is a continuous map, then a simplicial map $g : \Delta \longrightarrow \Gamma$ is called a *simplicial approximation* to f if $g(x) \in$ vert(carr$(f(x))$) for each $x \in |\Delta|$.

Proposition B.69. *If g is a simplicial approximation to f, then $f \simeq g$.*

Proof. The condition $g(x) \in$ vert(carr$(f(x))$) for each $x \in |\Delta|$ yields that the line segment $\{tg(x) + (1 - t)f(x) : t \in [0, 1]\}$ lies completely within the simplex carr$(f(x))$. Therefore the map

$$h : |\Delta| \times [0, 1] \longrightarrow |\Gamma|,$$
$$(x, t) \longmapsto tg(x) + (1 - t)f(x),$$

defines a homotopy from f to g. $\square$

The following *simplicial approximation theorem* makes simplicial homology theory work.

Theorem B.70. *Let $f : |\Delta| \longrightarrow |\Gamma|$ be a continuous map. Then for some $k \geq 0$, there exists a simplicial approximation $g :$ sd$^k \Delta \longrightarrow \Gamma$ to f.* $\square$

Simplicial Homology Theory

Consider the category of pairs of topological spaces that are polyhedra of some simplicial complex—in short *polyhedral spaces*—together with continuous maps.

For such a pair $(X, A) = (|\Delta|, |\Gamma|)$, let $K = K(\Delta)$ and $L = K(\Gamma)$ be the associated abstract complexes, and define the homology functor by $H_p(X, A) = H_p(K, L)$.

Any continuous map $f : (X, A) \longrightarrow (X', A')$ induces a map in homology by the simplicial approximation theorem (Theorem B.70) and Theorem B.65.

Let the family of natural transformations $\{\partial_p : p \in \mathbb{Z}\}$ be defined as on page 190.

The previous definitions constitute the *simplicial homology theory.* A proof of all the axioms is beyond the scope of this book.

Euler Characteristic

If Δ is a simplicial complex of dimension n, then define its *Euler characteristic* by

$$\chi(\Delta) = \sum_{i=0}^{n}(-1)^i \dim(H_i(K(\Delta); F)),$$

where F is an arbitrary field. This definition is consistent with the definition we gave on page 173 in the case that Δ is the triangulation of an orientable surface, since the simplicial homology for an orientable surface S of genus g easily computes to

$$H_p(S; F) = \begin{cases} F, & \text{if } p = 0 \text{ or } p = 2, \\ F^{2g}, & \text{if } p = 1, \\ 0, & \text{otherwise.} \end{cases}$$

The following remarkable *Euler–Poincaré theorem* holds.

Theorem B.71. *Let Δ be a simplicial complex of dimension n and let f_i be the number of simplices of dimension i, $i = 0, \ldots, n$. Then*

$$\chi(\Delta) = \sum_{i=0}^{n}(-1)^i f_i = \sum_{i=0}^{n}(-1)^i \dim C_i(K).$$
□

The Lefshetz–Hopf Fixed-Point Theorem

Let K be a simplicial complex, $f : K \longrightarrow K$ a simplicial map, and F a field. Then f induces F-linear maps $f_\Delta : C_*(K; F) \longrightarrow C_*(K; F)$ and $f_* : H_*(K; F) \longrightarrow H_*(K; F)$.

We want to relate the traces of the respective maps. If $g : V \longrightarrow V$ is an F-linear map of F-vector spaces, then denote its *trace* by $\operatorname{tr}(g)$; it is the sum of the diagonal entries of a matrix representation of g with respect to the same base of V.

The following *Hopf trace formula* achieves a relation between the traces of f_Δ and f_*. We will use the following abbreviations:

$$\operatorname{tr}_i(f_\Delta) = \operatorname{tr}(f_\Delta : C_i(K; F) \to C_i(K; F)),$$
$$\operatorname{tr}_i(f_*) = \operatorname{tr}(f_* : H_i(K; F) \to H_i(K; F)).$$

Theorem B.72. *Let K be a simplicial complex of dimension n, $f : K \longrightarrow K$ a simplicial map, and F a field. Then*

$$\sum_{i=0}^{n}(-1)^i \operatorname{tr}_i(f_\Delta) = \sum_{i=0}^{n}(-1)^i \operatorname{tr}_i(f_*).$$ □

Now, if Δ is a geometric simplicial complex of dimension n and $f : |\Delta| \longrightarrow |\Delta|$ a continuous map, then f also induces a map $f_* : H_*(|\Delta|; \Gamma) \longrightarrow H_*(|\Delta|; \Gamma)$. In this case, the number $L(f) = \sum_{i=0}^{n}(-1)^i \operatorname{tr}_i(f_*)$ is called the *Lefshetz number* of f. It is not hard to prove the following *Lefshetz–Hopf fixed-point theorem*.

Theorem B.73. *Let Δ be a geometric simplicial complex of dimension n, $f : |\Delta| \longrightarrow |\Delta|$ a continuous map, and F a field. If the Lefshetz number is nonzero, then f has a fixed point.*   □

Exercises

1. Show that $(X, \mathcal{O})$ is a topological space if and only if the following three conditions are satisfied:

   (a) $\emptyset, X \in \mathcal{O}$.
   (b) any *finite union* $A_1 \cup \cdots \cup A_k$ of closed sets $A_1, \ldots, A_k$ is closed.
   (c) any *arbitrary intersection* $\bigcap \mathcal{A}$ of closed sets $\mathcal{A}$ is closed.

2. Give a proof of Proposition B.4.
3. Let $(X, \mathcal{O})$ be a topological space and $A \subseteq X$. Show that int(A) is an open set contained in A and that it is inclusion maximal with this property.
4. Let $(X, \mathcal{O})$ be a topological space and $A \subseteq X$. Show that cl(A) is a closed set containing A and that it is inclusion minimal with this property.
5. Prove or disprove the following statements about a topological space $(X, \mathcal{O})$:

   (a) Each subset $A \subseteq X$ is either open or closed.
   (b) Any subset $U \subseteq X$ is open if and only if $U = \operatorname{int}(U)$.
   (c) Any subset $A \subseteq X$ is closed if and only if $A = \operatorname{cl}(A)$.
   (d) If $U \subseteq X$ is open, then $U \cap \partial U = \emptyset$.
   (e) If $A \subseteq X$ is closed and $B \subseteq A$, then $\operatorname{cl}(B) \subseteq A$.
   (f) If $U \subseteq X$ is open and $B \subseteq U$, then $\operatorname{cl}(B) \subseteq U$.
   (g) The boundary ∂A of any subset $A \subset X$ is closed.
   (h) cl(A) is the intersection of all closed sets containing A.
   (i) int(A) is the union of all open sets contained in A.

6. Give an example of a bijective continuous map such that the inverse map is not continuous.
7. Let $(X, \mathcal{O})$ be a topological space and $Y \subseteq X$.

   (a) Prove that the subspace topology $\mathcal{O}|_Y$ indeed defines a topology on Y.
   (b) Show that the subspace topology is the coarsest topology on Y such that the inclusion map $i : Y \hookrightarrow X$ is continuous.

8. Let $\{(X_\alpha, \mathcal{O}_\alpha) : \alpha \in A\}$ be an indexed family of topological spaces and $X = \coprod_{\alpha \in A} X_\alpha$ the set sum. Show that the family $\mathcal{O}_X$, as defined in Definition B.12, indeed defines a topology on X. Furthermore, show that $\mathcal{O}_X$ is the uniquely defined topology on X such that the following *universal property* holds: If Y is a topological space, then a map $g : X \longrightarrow Y$ is continuous if and only if all compositions $g \circ i_\alpha : X_\alpha \longrightarrow Y$ are continuous:

We say that $\mathcal{O}_X$ is the *final topology* of X with respect to the maps $(i_\alpha)_{\alpha \in A}$.

9. Let $\{(X_\alpha, \mathcal{O}_\alpha) : \alpha \in A\}$ be an indexed family of topological spaces and $X = \prod_{\alpha \in A} X_\alpha$ the product. Show that the family $\mathcal{O}_X$, as defined in Definition B.13, indeed defines a topology on X. Furthermore, show that $\mathcal{O}_X$ is the uniquely defined topology on X such that the following *universal property* holds: If Y is a topological space, then a map $g : Y \longrightarrow X$ is continuous if and only if all compositions $\pi_\alpha \circ g : Y \longrightarrow X_\alpha$ are continuous:

We say that $\mathcal{O}_X$ is the *initial topology* of X with respect to the maps $(\pi_\alpha)_{\alpha \in A}$.

10. Show that the definition of a join for subspaces of Euclidean space on page 168 and the subsequent definition for general topological spaces yield homeomorphic spaces in the case that $X \subseteq \mathbb{R}^m$ and $Y \subseteq \mathbb{R}^n$ are compact spaces.

11. A collection of sets has the *finite intersection property* if every intersection of a finite subcollection is nonempty. Show that a topological space X is compact if and only if for each collection $\mathcal{A}$ of closed subsets satisfying the finite intersection property, the intersection $\bigcap \mathcal{A}$ is nonempty.

12. Provide a proof of Proposition B.15 on page 169.

13. Provide a proof of Proposition B.16 on page 169.

14. Provide a proof of Proposition B.17 on page 169.

15. Show that the unit interval $I = [0, 1]$ is compact.

16. Let $X \subseteq \mathbb{R}^n$ be a compact subspace, $(A_k)_{k \geq 1}$ a sequence of closed subsets $A_k \subseteq X$ such that $\lim_{k \to \infty} \operatorname{diam}(A_k) = 0$. Show that there exists a point $x \in X$ such that each neighborhood of x contains infinitely many of the sets $A_k, k \geq 1$.

17. Let X be a compact metric space and $\mathcal{U}$ an open cover of X. Prove the existence of a *Lebesgue number,* i.e., the existence of a number $\lambda > 0$ such that for each $x \in X$, there exists a $U \in \mathcal{U}$ containing the λ-ball about x.

18. Consider the *configuration space* $L = \{\{x, y\} : x, y \in \mathbb{S}^1\}$ of two points on the circle, which are allowed to coincide. Its topology is defined to be the quotient topology with respect to the quotient map

$$\pi : \mathbb{S}^1 \times \mathbb{S}^1 \to L,$$

$$(x, y) \mapsto \{x, y\}.$$

Show that L is homeomorphic to the Möbius strip.

19. Show how a projective plane may be obtained from a polygon by a pairwise identification of directed edges.

20. Show that a Klein bottle may be obtained by gluing two Möbius strips along their common 1-dimensional sphere boundary.

21. Consider the quotient space

$$B = (\mathbb{S}^l \times [0, 1] \cup \mathbb{B}^{l+1})/ \sim,$$

where $\sim$ is the equivalence relation generated by $(x, 1) \sim x$. Prove that there is a homeomorphism $\varphi : B \to \mathbb{B}^{l+1}$ such that $\varphi(x, 0) = x$.

22. Let $f : \mathbb{S}^l \to X$ be a continuous map. Assume that f is homotopic to a map $g : \mathbb{S}^l \to X$ which extends to a map $G : \mathbb{B}^{l+1} \to X$. Prove that f extends to a map $F : \mathbb{B}^{l+1} \to X$.

23. Let $g : \mathbb{S}^l \to \mathbb{S}^n$ be a continuous map that is not surjective, $n \geq 0, l \geq -1$. Prove that g can be extended to a map $G : \mathbb{B}^{l+1} \to \mathbb{S}^n$.

24. Prove Theorem B.27, in other words, prove that for any $-1 \leq l \leq n - 1$, any continuous $f : \mathbb{S}^l \to \mathbb{S}^n$ can be extended to the ball $\mathbb{B}^{l+1}$. Hint: Use the previous exercises and Theorem B.70.

25. Apply Brouwer's fixed-point theorem, Theorem 1.1, in order to prove Theorem B.29 on page 171.

26. Prove that $x_0, \ldots, x_n \in \mathbb{R}^m$ are affinely independent if and only if for any $v = \sum_{i=0}^{n} \lambda_i x_i \in \mathrm{aff}(x_0, \ldots, x_n)$, the coefficients $\lambda_0, \ldots, \lambda_n$ are uniquely determined.

27. Prove that $x_0, \ldots, x_n \in \mathbb{R}^m$ are affinely independent if and only if the following implication holds for any set $\lambda_0, \ldots, \lambda_n \in \mathbb{R}$:

$$\left\{ \sum_{i=0}^{n} \lambda_i x_i = 0 \quad \text{and} \quad \sum_{i=0}^{n} \lambda_i = 0 \right\} \implies \lambda_0 = \cdots = \lambda_n = 0.$$

28. Let $i : \mathbb{R}^m \hookrightarrow \mathbb{R}^{m+1}$ be defined by $i(t_1, \ldots, t_m) = (1, t_1, \ldots, t_m)$. Show that $x_0, \ldots, x_n \in \mathbb{R}^m$ are affinely independent if and only if $i(x_0), \ldots, i(x_n) \in \mathbb{R}^{m+1}$ are linearly independent.

29. Let Δ be a geometric simplicial complex and $f : \mathrm{vert}(\Delta) \to \mathbb{R}^d$ any map. Show that f induces a unique map $\bar{f} : |K| \to \mathbb{R}^d$ that is affine linear on each simplex $\sigma \in \Delta$.

30. If you have not yet done so in your life, compute the Vandermonde determinant

$$\det \begin{pmatrix} 1 & t_0 & t_0^2 & \cdots & t_0^m \\ 1 & t_1 & t_0^2 & \cdots & t_1^m \\ & & \vdots & & \\ 1 & t_m & t_m^2 & \cdots & t_m^m \end{pmatrix}.$$

31. Give an alternative proof of Lemma B.40 along the following lines. To show that the vectors $x_0 = (1, t_0, \ldots, t_0^m), \ldots, x_m = (1, t_m, \ldots, t_m^m)$ are linearly independent, assume that for some j, the vector x_j is contained in the linear span $A = \langle \{ x_i : i \neq j \} \rangle$. Let $y \in \mathbb{R}^{m+1}$ be a nontrivial vector orthogonal to A. Now consider the polynomial map $t \mapsto \langle y, (1, t, \ldots, t^m)^t \rangle$.

32. Give a proof of Proposition B.42 on page 178.

33. Give a proof of Proposition B.46 on page 180.

34. Let Γ and Δ be geometric simplicial complexes in $\mathbb{R}^m$ and $\mathbb{R}^n$, respectively. Give a definition of their join, i.e., define a geometric simplicial complex $\Gamma * \Delta$ such that $K(\Gamma * \Delta) = K(\Gamma) * K(\Delta)$.

35. Prove the exactness of the long homology sequence in Theorem B.56 on page 185.

36. Give a proof of Theorem B.63 on page 188.

37. Give a proof of Proposition B.64 on page 189, i.e., show that ∂ respects the relations on the oriented simplices and satisfies $\partial^2 = 0$.

38. Let (K, L, M) be a *triple of simplicial complexes*, i.e., (K, L) and (L, M) are pairs of simplicial complexes. Show that there is a short exact sequence

$$0 \longrightarrow C_*(L, M) \xrightarrow{i} C_*(K, M) \xrightarrow{\pi} C_*(K, L) \longrightarrow 0$$

inducing the *long exact sequence of a triple*

$$\cdots \xrightarrow{\partial_{p+1}} H_p(L, M) \xrightarrow{i_*} H_p(K, M) \xrightarrow{\pi_*} H_p(K, L) \xrightarrow{\partial_p} H_{p-1}(L, M) \xrightarrow{i_*} \cdots .$$

Chapter 7
Appendix C: Partially Ordered Sets, Order Complexes, and Their Topology

Partially ordered sets and the interplay with their associated simplicial complexes, as well as the topology of these complexes, all play a substantial role in topological combinatorics. We will briefly introduce a few of the concepts.

C.1 Partially Ordered Sets

A *partially ordered set*, commonly abbreviated as a *poset,* is given by a pair $(P, \leq)$, where P is a set and $\leq$ is a binary relation on P satisfying, for any $x, y, z \in P$, the properties of being

- (Reflexive) $x \leq x$,
- (Transitive) $x \leq y$ and $y \leq z$ imply $x \leq z$, and
- (Antisymmetric) $x \leq y$ and $y \leq x$ imply $x = y$.

We will exclusively be concerned with the case in which P is finite, i.e., with finite posets. If no confusion may occur about the relation $\leq$, we might refer to P itself as a poset.

The most prominent example is given by the *Boolean poset, B_n*, consisting of all subsets of $[n] = \{1, \ldots, n\}$ ordered by inclusion $\subseteq$.

A finite poset is often represented by its *Hasse diagram*. Figure C.1 shows the Hasse diagram of B_3, which should suffice as an illustration of the concept.

Another important example is the *face poset, $\mathcal{F}(K)$*, of an abstract simplicial complex, K, which is defined by the set of all nonempty faces of K, i.e.,

$$\mathcal{F}(K) = \{\sigma \in K : \sigma \neq \emptyset\},$$

ordered by inclusion.

If $(P, \leq)$ is a poset and $Q \subseteq P$ is a subset, then the restriction of the relation $\leq$ to Q yields the *subposet* $(Q, \leq|_Q)$.

M. de Longueville, *A Course in Topological Combinatorics*, Universitext,
DOI 10.1007/978-1-4419-7910-0_7,
© Springer Science+Business Media New York 2013

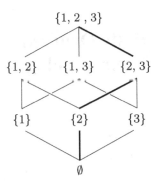

Fig. C.1 Hasse diagram of B_3

A poset $(P, \leq)$ is called *linearly ordered* if any two elements are comparable, i.e., for any $x, y \in P$, we have either $x \leq y$ or $y \leq x$. In a linearly ordered poset, the elements of P may be labeled in a linear fashion, i.e., we may write $P = \{p_1, \ldots, p_n\}$ such that $p_1 \leq p_2 \leq \cdots \leq p_n$.

A *chain* in a poset P is a subset $C \subseteq P$ such that as a subposet, C is linearly ordered. A chain is called *maximal* if it is inclusion maximal, i.e., it cannot be refined by adding elements. An example of a maximal chain in B_3 is given by $\{\emptyset, \{2\}, \{2,3\}, \{1,2,3\}\}$. It is highlighted in Fig. C.1 by the bold edges.

The *length* of a chain C is defined to be $|C| - 1$, i.e., the number of elements -1.

A poset P is *pure* if all maximal chains have the same length. It is *bounded* if it has a least element $\hat{0}$ and a greatest element $\hat{1}$, i.e., $\hat{0} \leq x \leq \hat{1}$ for all $x \in P$. Any finite poset P can be made into a bounded one by adding a least and a greatest element. We will denote this by $\hat{P} = P \cup \{\hat{0}, \hat{1}\}$. A finite, pure, and bounded poset P admits a *rank function*, $\rho : P \to \mathbb{N}$, that assigns to each element x the common length of all maximal chains from $\hat{0}$ to x.

If $x \leq y$ in a poset P, then denote by $[x, y] = \{z : x \leq z \leq y\}$ the *interval* of elements lying between x and y.

Poset Maps

Let $(P, \leq_P)$ and $(Q, \leq_Q)$ be two posets. A map $f : P \to Q$ is called *order-preserving (resp. -reversing)* if for any $x \leq_P y$ in P, we have $f(x) \leq_Q f(y)$ (resp. $f(y) \leq_Q f(x)$). Any such map will be referred to as a *poset map*. Note that poset maps, in particular, map chains to chains.

A *poset isomorphism* is a bijective poset map whose inverse is also a poset map.

Product Orders

Let $(P, \leq_P)$ and $(Q, \leq_Q)$ be two posets. The *product* $P \times Q$ becomes a poset via $(p_0, q_0) \leq (p_1, q_1)$ if and only if $p_0 \leq_P p_1$ and $q_0 \leq_Q q_1$.

Consider the following example. Let $I = \{0, 1\}$ be the poset defined by $0 < 1$. Let $P = I \times \cdots \times I$ be the *n*-fold product of I. We claim that P is isomorphic to the Boolean poset B_n. In fact, the map

$$f : P \longrightarrow B_n,$$

$$(t_1, \ldots, t_n) \longmapsto \{i \in [n] : t_i = 1\},$$

yields a poset isomorphism. The bijectivity is clear. The fact that f and its inverse are order-preserving is due to the following computation:

$$p = (t_1, \ldots, t_n) \leq (t_1', \ldots, t_n') = p' \iff \left(\forall i : t_i = 1 \implies t_i' = 1\right)$$

$$\iff \left(\forall i : i \in f(p) \implies i \in f(p')\right)$$

$$\iff f(p) \subseteq f(p').$$

Lexicographic Order

If $P_1, \ldots, P_n$ are linear orders, then the *lexicographic order* $\preceq$ is a linear order on the product $P = P_1 \times \cdots \times P_n$ (considered as a product of sets). It is defined as follows. Let $(p_1, \ldots, p_n), (q_1, \ldots, q_n) \in P$. Then $(p_1, \ldots, p_n) \preceq (q_1, \ldots, q_n)$ if either $(p_1, \ldots, p_n) = (q_1, \ldots, q_n)$ or there exists a $k \in \{1, \ldots, n\}$ such that $p_i = q_i$ for $1 \leq i < k$ and $p_k < q_k$ with respect to the order of P_k.

C.2 Order Complexes

To any partially ordered set $(P, \leq)$ we associate its *order complex,* $\Delta(P)$, which is an abstract simplicial complex whose simplices are given by chains in P, i.e.,

$$\Delta(P) = \{\{s_0, \ldots, s_k\} \subseteq P : s_0 \leq \cdots \leq s_k\}.$$

The facets of $\Delta(P)$ are given by the maximal chains of P. Hence if P is a pure poset, all facets of $\Delta(P)$ will have the same dimension. An example of a poset—given by its Hasse diagram—and its order complex is shown in Fig. C.2.

By definition of the barycentric subdivision of an abstract simplicial complex (given on page 179), we obtain the important proposition.

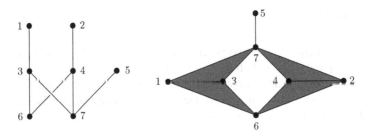

Fig. C.2 An example of a poset P and its order complex $\Delta(P)$

Proposition C.1. *The barycentric subdivision of an abstract simplicial complex K is identical to the order complex of the face poset of K, i.e.,* sd $K = \Delta(\mathcal{F}(K))$. □

Product Triangulation

Assume that $\Delta(P)$ and $\Delta(Q)$ are geometrically realized by maps $x : P \to \mathbb{R}^n$ and $y : Q \to \mathbb{R}^m$. We are interested in the product of polyhedra $|\Delta(P)| \times |\Delta(Q)|$. This product is naturally triangulated by the geometric realization of $\Delta(P \times Q)$ given by the map $x \times y : P \times Q \to \mathbb{R}^n \times \mathbb{R}^m$. In particular, we obtain the following result.

Lemma C.2. *If P and Q are finite posets, then there is a natural homeomorphism* $|\Delta(P \times Q)| \cong |\Delta(P)| \times |\Delta(Q)|$. □

Simplicial Maps of Order Complexes and Homotopy

Since a poset map $f : P \to Q$ maps chains to chains, it induces a simplicial map $f : \Delta(P) \to \Delta(Q)$ of the corresponding order complexes.

The concept of order complexes allows us to speak in topological terms about partially ordered sets. In particular, we will call two poset maps $f, g : P \to Q$ *homotopic*, denoted by $f \simeq g$, if the induced maps on the geometric realizations $|f|, |g| : |\Delta(P)| \to |\Delta(Q)|$ are homotopic. Correspondingly, we say that two posets, P and Q, are *homotopy equivalent*, denoted by $P \simeq Q$, if there exist poset maps $f : P \to Q$ and $g : Q \to P$ such that the compositions $g \circ f$ and $f \circ g$ are homotopic to the respective identity maps.

The following *order homotopy lemma* has quite strong applications.

Lemma C.3. *Let $f, g : P \to Q$ be two poset maps of finite posets such that $f \leq g$, i.e., $f(x) \leq g(x)$ for all $x \in P$. Then $f \simeq g$.*

Proof. Let the poset $I = \{0, 1\}$ be defined by $0 < 1$ and consider the product $P \times I$ endowed with the product order. Now define the map $h : P \times I \to Q$ by $h(x, 0) = f(x)$ and $h(x, 1) = g(x)$, which is clearly order-preserving. Then h induces a map

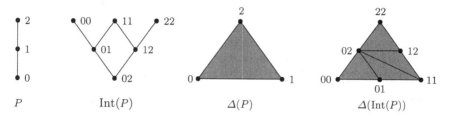

Fig. C.3 A poset P, its interval order $\mathrm{Int}(P)$, and the complexes $\Delta(P)$ and $\Delta(\mathrm{Int}(P))$

$|h| : |\Delta(P \times I)| \to |\Delta(Q)|$. By the previous lemma, $|\Delta(P \times I)| \cong |\Delta(P)| \times [0, 1]$, so we obtain a map $H : |\Delta(P)| \times [0, 1] \to |\Delta(Q)|$. It is now easy to check that $H(x, 0) = |f|(x)$ and $H(x, 1) = |g|(x)$. $\qquad\qquad\square$

Remark C.4. Note that if f and g are equivariant with respect to some action of a group G on P and Q, then so is the homotopy H.

Interval Order

If $(P, \leq)$ is a poset, we may consider the set of all pairs (x, y) with $x \leq y$ corresponding to intervals of P. An interval defined by (x, y) contains an interval defined by (x', y') if and only if $x \leq x'$ and $y' \leq y$. We use this interpretation to define the poset $\mathrm{Int}(P) = \{(x, y) : x \leq y\}$ ordered by

$$(x, y) \preceq (x', y') \iff x \leq x' \text{ and } y' \leq y,$$

and call it the *interval order of* P. An illustration is given in Fig. C.3 with the abbreviations xy for (x, y).

Apparently, the two posets are very closely related. James Walker [Wal83] has proved the following result.

Proposition C.5. *The geometric realizations $|\Delta(P)|$ and $|\Delta(\mathrm{Int}(P))|$ are naturally homeomorphic.* $\qquad\qquad\square$

C.3 Shellability of Partial Orders

A finite poset P is called *(pure) shellable* if it is pure and its maximal chains can be ordered $m_1, \ldots, m_t$ in such a way that if $1 \leq i < j \leq t$, then there exist $1 \leq k < j$ and $x \in m_j$ such that $m_i \cap m_j \subseteq m_k \cap m_j = m_j \setminus \{x\}$. In other words, the intersection of a chain m_j with any preceding chain m_i is contained in an intersection $m_k \cap m_j$ of maximal size for some other chain m_k preceding m_j.

The concept was introduced by Björner [Bjö80] and investigated extensively by Björner and Wachs [BW83]. Its importance becomes clear by its connection to the topological concept of shellability, which we discussed on page 178.

Proposition C.6. *If P is shellable, then the associated order complex $\Delta(P)$ is shellable.*

Proof. The proof is an easy exercise. □

Since our definition of shellability requires the poset P to be pure, we obtain the following immediate corollary.

Corollary C.7. *If P is shellable, then $|\Delta(P)|$ is, up to homotopy, either contractible or a wedge of spheres of dimension* $\dim(\Delta)$. □

Quite some theory has been developed about special types of shellability. Here we want to exemplify the concept of a lexicographic shelling for the Boolean poset.

Proposition C.8. *The Boolean poset B_n is shellable.*

Proof. First of all, we will be concerned with associating label vectors to maximal chains c in any interval $[a, b]$ in B_n. If c is given by the chain $a = x_0 \subset x_1 \subset \cdots \subset x_k = b$, then any two successive elements in this sequence differ by an element of $[n]$, say $x_{j+1} \setminus x_j = \{i_{j+1}\}$. The label vector we associate with c is now defined to be the vector $\lambda(c) = (i_1, \ldots, i_k)$. Hence the label vectors for the maximal chains m of B_n are given by vectors $\lambda(m) \in [n]^n$. For example, the maximal chain shown in bold in Fig. C.1 obtains the label vector $(2, 3, 1)$.

We now order the maximal chains $m_1, \ldots, m_t$ according to the lexicographic order of their label vectors, i.e., $\lambda(m_1) \prec \lambda(m_2) \prec \cdots \prec \lambda(m_t)$. Note that if $[a, b]$ is an interval in B_n with $b \setminus a = \{i_1, \ldots, i_k\}$ with $i_1 < i_2 < \cdots < i_k$, then there is a unique maximal chain c in $[a, b]$ with *increasing* label vector $(i_1, \ldots, i_k)$, and this label vector is the lexicographically smallest among all label vectors of maximal chains in $[a, b]$.

Now let $1 \le i < j \le t$, and consider the chains m_i and m_j given by $\emptyset = y_0 \subset y_1 \subset \cdots \subset y_n = [n]$ and $\emptyset = x_0 \subset x_1 \subset \cdots \subset x_n = [n]$, respectively. Let r be maximal with the property that $x_i = y_i$ for all $0 \le i \le r$, and let $s > r$ be minimal with $x_s = y_s$. In other words, the chains m_i and m_j agree all the way up to $x_r = y_r$ and afterward meet for the first time at $x_s = y_s$. The situation is sketched in Fig. C.4.

Since $\lambda(m_i) \prec \lambda(m_j)$, the unique chain c in the interval $[x_r, x_s]$ with increasing label vector will certainly be different from $x_r \subset x_{r+1} \subset \cdots \subset x_s$. Hence in the corresponding label vector, $\lambda(x_r \subset x_{r+1} \subset \cdots \subset x_s)$, there must exist a descent, i.e., there exist $r < t < s$ such that $\alpha = \lambda(x_{t-1} \subset x_t) > \lambda(x_t \subset x_{t+1}) = \beta$. In other words, $x_t = x_{t-1} \cup \{\alpha\}$, $x_{t+1} = x_t \cup \{\beta\}$, and $\alpha > \beta$; see Fig. C.5. If we set $x = x_{t-1} \cup \{\beta\}$, then clearly $x_{t-1} \subset x \subset x_{t+1}$, and the maximal chain $x_0 \subset \cdots \subset x_{t-1} \subset x \subset x_{t+1} \subset \cdots \subset x_n$ precedes m_j with respect to the

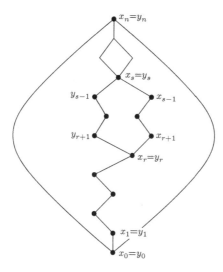

Fig. C.4 The definition of r and s

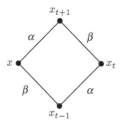

Fig. C.5 The descent at x_t and the element x

lexicographic order of its label vector. Hence this chain corresponds to a chain m_k for some $k < j$. Moreover, $m_i \cap m_j = m_i \cap m_k \subseteq m_k \cap m_j = m_j \setminus \{x\}$. □

Now let P be a shellable poset. If P is bounded, let ρ be the rank function of P; otherwise, let ρ be the rank function of $\hat{P}$ and assume that $\rho(\hat{1}) = r + 1$. For $S \subseteq [r]$, let $P_S = \{x \in P : \rho(x) \in S\}$ be the *rank selected subposet of P defined by S*.

Theorem C.9. *If P is shellable, then P_S is shellable for any $S \subseteq [r]$.*

Proof. Let $m_1, \ldots, m_t$ be a shelling order of P. For each maximal chain c of P_S, let $\theta(c) = \min\{i : c \subseteq m_i\}$ be the minimal index i such that $c \subseteq m_i$. Clearly θ is an injective map on the set of all maximal chains of P_S. Hence we may define a linear order $\lhd$ on the set of maximal chains of P_S by $c \lhd d$ if $\theta(c) < \theta(d)$. Now assume

$c \lhd d$. We have to find a maximal chain $e \lhd d$ of P_S such that $c \cap d \subseteq e \cap d = d \setminus \{x\}$ for some $x \in d$. But since $m_1, \ldots, m_t$ is a shelling order, there exists a $k < \theta(d)$ such that

$$m_{\theta(c)} \cap m_{\theta(d)} \subseteq m_k \cap m_{\theta(d)} = m_{\theta(d)} \setminus \{x\} \tag{$*$}$$

for some $x \in m_{\theta(d)}$. The chain $e = m_k \cap P_S$ is clearly maximal in P_S, and since $\theta(e) \le k < \theta(d)$, we have $e \lhd d$. Assuming that $x \notin d$, we obtain

$$d \subseteq m_{\theta(d)} \setminus \{x\} = m_k \cap m_{\theta(d)} \subseteq m_k,$$

contradicting the fact that $\theta(d)$ is minimal among all i with $d \subseteq m_i$. Hence $x \in d$, and by intersecting the relation $(*)$ with P_S, we obtain $c \cap d \subseteq e \cap d = d \setminus \{x\}$ as desired. $\square$

Corollary C.10. *The rank selected subposet*

$$B_{n,k} = \{S \subseteq [n] : k \le |S| \le n - k\}$$

of the Boolean poset B_n is shellable. $\square$

Exercises

1. Let $(P, \le)$ be a partially ordered set. Show that there exists a linear order $\preceq$ on P such that the identity map from P to itself is an order-preserving map from $(P, \le)$ to $(P, \preceq)$. The order $(P, \preceq)$ is called a *linear extension of* $(P, \le)$.
2. Let $(P, \le)$ be a finite partially ordered set. Show that there exist finitely many linear extensions $(P, \preceq_1), \ldots, (P, \preceq_n)$ of $(P, \le)$ such that $x \le y$ if and only if for all $i = 1, \ldots, n$, the relation $x \preceq_i y$ holds.
3. Let $(P, \le)$ be a finite partially ordered set. A subset $A \subseteq P$ is called an *antichain* if its elements are pairwise incomparable. Let h denote the largest cardinality of a chain in P. Prove that it is always possible to partition P into h antichains.
4. Provide the details for the proof of Lemma C.2.
5. Provide a proof of Proposition C.6.
6. The labeling in the proof of Proposition C.8 may be thought of as induced by a labeling of the edges of the Hasse diagram of the Boolean poset. We may use this idea to obtain a general concept of an *EL-labeling*, i.e., an *edge lexicographic labeling* of a poset.

    Let $\lambda : \mathcal{E}(P) \to \mathbb{Z}$ be a *labeling of the edges* $\mathcal{E}(P)$ of the Hasse diagram of P. If c is a maximal chain, $a = x_0 < x_1 < \cdots < x_k = b$, in some interval $[a, b]$ of P, then we may associate the label vector $\lambda(c) = (\lambda(x_0, x_1), \ldots, \lambda(x_{k-1}, x_k))$ with c. Such a chain c is called *increasing* if $\lambda(x_0, x_1) < \cdots < \lambda(x_{k-1}, x_k)$.

An edge labeling λ is defined to be an *EL-labeling* if for every interval $[a, b]$ in P,

(a) There is a unique increasing maximal chain c in $[a, b]$, and
(b) $\lambda(c) \prec \lambda(c')$ in the lexicographic order for all other maximal chains c' in $[a, b]$.

Prove along the lines of the proof of Proposition C.8 that any poset P that admits an EL-labeling is shellable.

Chapter 8
Appendix D: Groups and Group Actions

Historically, the study of symmetries of geometric bodies led to the concept of a group. If one considers a point x of a geometric body, then a symmetry g of the body will map x to some point that we denote by $g \circ x$. We say that the group acts on the set of points of the body. In this chapter we briefly recall the fundamental concepts of groups and their actions on spaces.

D.1 Groups

A *group* is given by a triple $(G, \cdot, e)$, where G is a nonempty set, $\cdot$ is a binary operation $\cdot : G \times G \to G$, also called multiplication, and $e \in G$ is the *neutral element* subject to the following axioms:

- (Associativity) $(a \cdot b) \cdot c = a \cdot (b \cdot c)$ holds for all $a, b, c \in G$.
- (Neutral element) $g \cdot e = e \cdot g = g$ for all $g \in G$.
- (Inverses) For every $g \in G$ there exists *an inverse element*, denoted by g^{-1}, such that $g \cdot g^{-1} = g^{-1} \cdot g = e$.

In practice, the multiplication symbol is often suppressed, and we write ab instead of $a \cdot b$. If no confusion will occur, we refer to the group just by G. A group G is called *abelian (or commutative)* if $ab = ba$ for all $a, b \in G$. The binary operation on an abelian group is usually called addition and denoted by $+$.

Some prominent examples of (abelian) groups are the integers $(\mathbb{Z}, +, 0)$, the integers modulo some fixed integer $(\mathbb{Z}_m, +, 0)$, powers $((\mathbb{Z}_m)^r, +, 0)$ thereof with componentwise addition, and the nonzero reals $(\mathbb{R} \setminus \{0\}, \cdot, 1)$. An important example of a nonabelian group is given by the *symmetric group* $(\operatorname{Sym}(X), \circ, \operatorname{id}_X)$, i.e., the bijections of a (finite) set X together with composition.

M. de Longueville, *A Course in Topological Combinatorics*, Universitext,
DOI 10.1007/978-1-4419-7910-0_8,
© Springer Science+Business Media New York 2013

Subgroups and Homomorphisms

If $(G, \cdot, e)$ is a group, then H is a *subgroup of* G, denoted by $H \leq G$, if $H \subseteq G$ and $(H, \cdot|_{H \times H}, e)$ is a group. A subgroup is called *normal*, denoted by $H \trianglelefteq G$, if $aHa^{-1} \subseteq H$ for all $a \in G$.

An important (abelian) subgroup of any group G is its *center* $C(G) = \{a \in G : ax = xa$ for all $x \in G\}$.

A map $f : G \rightarrow H$ of groups is a *homomorphism* if $f(ab) = f(a)f(b)$ for all $a, b \in G$. The *kernel of a homomorphism* f is defined to be the preimage of the neutral element $\ker(f) = f^{-1}(e)$. The kernel of a homomorphism is a normal subgroup, and conversely, each normal subgroup can be realized as the kernel of a homomorphism via the following concept.

If $H \trianglelefteq G$ is a normal subgroup, then the *quotient group* G/H is defined to be the set of *cosets* $\{gH : g \in G\}$ with the multiplication $aH \cdot bH = abH$ and neutral element $eH = H$. There is a natural *quotient homomorphism* $\pi : G \rightarrow G/H$ defined by $g \mapsto gH$ whose kernel is $\ker(\pi) = H$.

The *order of a group* G is defined to be the number $|G|$ of its elements. The *order of an element* $a \in G$ is defined to be the order of the subgroup generated by a, i.e., in the case of a finite order, the smallest $k \geq 1$ such that $a^k = e$. The *index* $[G : H]$ *of a subgroup* $H \leq G$ is defined to be the number of cosets $\{gH : g \in G\}$. The following result is attributed to Joseph Louis Lagrange.

Proposition D.1. *If $H \leq G$ is a subgroup, then $|G| = [G : H]|H|$. In particular, if G is of finite order, then $|H|$ is a divisor of $|G|$.*   □

A group G is called *cyclic* if there exists an element a generating the whole group, i.e., $G = \{a^n : n \in \mathbb{Z}\}$.

If p is a prime number, then a finite group G is called a *p-group* if the order of the group is a power of p. In other words, G is a p-group if $|G| = p^k$ for some $k \geq 1$.

D.2 Group Actions

We will now introduce the general idea of a group acting on a set. It may be helpful, though, to think of a geometric body and its symmetries as explained in the beginning of the chapter.

Definition D.2. Let $(G, \cdot, e)$ be a group and X a set. An *action of G on X* is given by a map

$$\circ : G \times X \longrightarrow X,$$

$$(g, x) \longmapsto g \circ x,$$

subject to the following two conditions:

$$e \circ x = x,$$

$$g \circ (h \circ x) = (g \cdot h) \circ x,$$

where $x \in X$, e is the neutral element of G, and $g, h \in G$.

If we consider an action of a group G on a set X, we also say that G *acts on* X. The identity

$$(g^{-1}) \circ (g \circ x) = (g^{-1} \cdot g) \circ x = e \circ x = x$$

shows that any element $g \in G$ induces a bijection $g \circ - : X \to X$. In fact, a group action induces—and may be defined by—a group homomorphism

$$G \longrightarrow \mathrm{Sym}(X),$$

$$g \longmapsto g \circ -,$$

from G to the symmetric group of X as discussed in the exercises. Therefore, we think of the elements of G as symmetries of the set X.

If no confusion may arise, we will drop the symbol $\circ$ and write gx instead of $g \circ x$.

Let us consider a few examples. Let G be any group. Then, taking $X = G$, the group G acts on itself by $g \circ h = gh$. For any group G and any set X there is always the *trivial action* defined by $g \circ x = x$ for all $g \in G$ and $x \in X$. A more concrete example is the following. Let $G = \mathbb{Z}_2 = \{e, \tau\}$ be the 2-element group and $X = \mathbb{Z}$ the integers. Then a group action of G on X is defined by $\tau \circ m = -m$.

Equivariant Maps

If G acts on sets X and Y, then a map $f : X \to Y$ is called G-*equivariant, or equivariant with respect to G*, if $f(gx) = gf(x)$ for all $g \in G$ and $x \in X$.

An instructive example that occurs in similar guise in this book is the following. Let $f : A \to B$ by any map of sets A and B. Let $X = A \times \cdots \times A$ and $Y = B \times \cdots \times B$ be the k-fold Cartesian products for some $k \geq 1$. The symmetric group $G = \mathrm{Sym}(k)$ acts on X and Y by permuting the coordinates, i.e., $\pi(x_1, \ldots, x_k) = (x_{\pi(1)}, \ldots, x_{\pi(k)})$. Then the map $F = f^k : X \to Y$ defined by $F(x_1, \ldots, x_k) = (f(x_1), \ldots, f(x_k))$ is clearly G-equivariant.

Orbits and Fixed Points

Definition D.3. If G acts on the set X, then for $x \in X$, the set $G \circ x = \{g \circ x : g \in G\}$ is called *the orbit of* x. The set of orbits is denoted by X/G. A group action is called *transitive* if there exists only one orbit, i.e., $G \circ x = X$ for one (and hence for all) $x \in X$.

Clearly the orbits yield a partition of X. In our previous examples, the action of a group on itself is transitive and hence has only the orbit $X = G$, the trivial action has all singletons $\{x\}$, for $x \in X$, as orbits; and the orbits of the third example are the sets $\{0\}$ and $\{m, -m\}$ for $m \in \mathbb{Z}$.

Definition D.4. If G acts on the set X, then the *fixed-point set of the group action* is defined to be

$$X^G = \{x \in X : \forall g \in G : g \circ x = x\},$$

i.e., the set of all points that are simultaneously fixed by all group elements. The group action is called *fixed-point-free* if $X^G = \emptyset$.

The action of a group on itself is fixed-point-free, the trivial action has $X^G = X$ by definition, and in the third example only the element 0 is fixed by all group elements; hence $X^G = \{0\}$ in this case.

Induced Actions

If G acts on X via $\circ$ and $H \leq G$ is a subgroup, then clearly H acts on X via

$$H \times X \longrightarrow X,$$
$$(h, x) \longmapsto h \circ x.$$

If, moreover, $H \trianglelefteq G$ is a normal subgroup, then the quotient G/H acts on X^H via

$$G/H \times X^H \longrightarrow X^H,$$
$$(gH, x) \longmapsto g \circ x.$$

Lemma D.5. *If* $H \trianglelefteq G$ *is a normal subgroup, then* $X^G = (X^H)^{G/H}$. $\qquad\qquad \square$

Stabilizer Subgroup

Definition D.6. If G acts on X and $x \in X$, then the *stabilizer (or isotropy) subgroup G_x of x* is defined by

$$G_x = \{g \in G : g \circ x = x\}.$$

A group action is called *free* if $G_x = \{e\}$ is trivial for every $x \in X$. In other words, for every $g \in G$ and $x \in X$, $g \circ x = x$ implies that $g = e$ is the neutral element.

Lemma D.7. *If a finite group G acts on X and $x \in X$, then $|G \circ x| = |G|/|G_x|$.*

Proof. Consider the map

$$\varphi : G \longrightarrow G \circ x,$$

$$g \longmapsto g \circ x.$$

Then $\varphi(g) = \varphi(h)$ if and only if $(g^{-1}h) \circ x = x$ if and only if $h \in gG_x$. Hence $\varphi^{-1}(\varphi(g)) = gG_x$. Now clearly φ is surjective and $|gG_x| = |G_x|$. $\qquad\square$

In particular, this implies that if G acts freely, then the size of each orbit is the order of the group G.

The Class Equation and p-Groups

Let G be a finite group. Define the *conjugacy action* of G on $X = G$ via $g \circ x = gxg^{-1}$. The orbits $G \circ x = \{gxg^{-1} : g \in G\}$ are called the *conjugacy classes* of G. In this case the stabilizer subgroup $G_x = \{g \in G : gxg^{-1} = x\}$ is called the *centralizer of x*. Note that $x \in G$ is an element of the center $C(G)$ if and only if its centralizer G_x is equal to G. Phrased differently, $x \in C(G)$ if and only if $[G : G_x] = 1$.

Now let $G \circ x_1, \ldots, G \circ x_n$ be the distinct orbits of the G-action. Then clearly

$$|G| = \sum_{i=1}^{n} |G \circ x_i| = \sum_{i=1}^{n} |G|/|G_{x_i}| = \sum_{i=1}^{n} [G : G_{x_i}].$$

Without loss of generality we may order the x_i such that $x_1, \ldots, x_m \in G \setminus C(G)$ and $x_{m+1}, \ldots, x_n \in C(G)$. We obtain the *class equation*

$$|G| = |C(G)| + \sum_{i=1}^{m} [G : G_{x_i}],$$

where $[G : G_{x_i}] > 1$ for all $i = 1,\ldots,m$. We will apply the class equation in order to obtain a result on p-groups.

Proposition D.8. *If G is a p-group of order p^k for some $k \geq 1$, then there exists a sequence of normal subgroups*

$$\{e\} = G_0 \subseteq G_1 \subseteq \cdots \subseteq G_k = G$$

such that G_{i+1}/G_i is cyclic of order p.

Proof. In the class equation, $|G|$ and each index $[G : G_{x_i}]$, $i = 1,\ldots,m$, are divisible by p. Hence $|C(G)|$ must be divisible by p as well. In particular, $C(G) \neq \{e\}$ is nontrivial. We now proceed by induction on k. The case $k = 1$ is clear. For the induction step, assume that $k > 1$. Then there exists an element $a \in C(G)$ of order p. Such an element is easily found: since $C(G)$ is nontrivial, let $b \in C(G) \setminus \{e\}$ be an arbitrary element. Its order will be some power of p, say p^l. Then set $a = b^{p^{l-1}}$. Let G_1 be the subgroup generated by a, i.e., $G = \{e,a,a^2,\ldots,a^{p-1}\}$. Clearly $G_1 \subseteq C(G)$, and hence G_1 is normal in G. Set $H = G/G_1$. Then H has order p^{k-1} by Proposition D.1, and by the induction hypothesis, it has a sequence of normal subgroups

$$\{e\} = H_1 \subseteq H_2 \subseteq \cdots \subseteq H_k = H$$

such that H_{i+1}/H_i is cyclic of order p. For $i = 2,\ldots,k$ set $G_i = \pi^{-1}(H_i)$, where $\pi : G \to G/G_1$ denotes the natural quotient map.   □

D.3 Topological G-Spaces

If X is a topological space, then X is called a *G-space* if G acts continuously on X, i.e., for each $g \in G$, the map $g \circ -$ is continuous. If, moreover, the action is free, X is called a *free G-space*.

The most prominent example of a free G-space in this book is the n-dimensional sphere with the antipodal action, i.e., $G = \mathbb{Z}_2 = \{e, \nu\}$, $X = \mathbb{S}^n$ with $\nu \circ x = -x$.

If X is a G-space, then consider the natural quotient map to the *orbit space:*

$$\pi : X \longrightarrow X/G,$$

$$x \longmapsto Gx.$$

It induces the quotient topology on X/G, i.e., $U \subseteq X/G$ is defined to be open if $\pi^{-1}(U) \subseteq X$ is open.

An illustrative example is the following. Let $G = \mathbb{Z}_n$ and let $X = \mathbb{S}^1$ be the 1-dimensional sphere. Then X becomes a G-space via $[k]\, e^{i\varphi} = e^{i(\varphi + k\frac{2\pi}{n})}$. Then each orbit looks like $\mathbb{Z}_n \circ e^{i\varphi} = \{e^{i\varphi}, e^{i(\varphi + \frac{2\pi}{n})}, e^{i(\varphi + 2\frac{2\pi}{n})}, \ldots, e^{i(\varphi + (n-1)\frac{2\pi}{n})}\}$, and the orbit space is homeomorphic to a 1-dimensional sphere itself via

$$X/G \longrightarrow \mathbb{S}^1,$$

$$\mathbb{Z}_n \circ e^{i\varphi} \longmapsto e^{in\varphi}.$$

Equivariant Maps and Homotopies

A G-equivariant map $f : X \to Y$ of topological G-spaces will always be assumed continuous.

If $f, g : X \to Y$ are G-equivariant maps, then a *G-equivariant homotopy from f to g* is a homotopy $H : X \times [0, 1] \to Y$ from f to g with the additional property that for each $t \in [0, 1]$, the map $H(-, t)$ is G-equivariant.

D.4   Simplicial Group Actions

Let G be a finite group and K a simplicial complex. A group action $G \times \mathrm{vert}(K) \to \mathrm{vert}(K)$ of G on K is *simplicial* if for each $g \in G$ the map $g \circ -$ is a simplicial map, in other words, if the action induces an action of G on K itself. A simplicial complex K with a simplicial G-action is called a *G-complex*. Moreover, if the induced action of G on K is free, then K is called a *free G-complex*.

A simplicial group action of G on K induces a continuous group action on the geometric realization $|K|$, thus turning $|K|$ into a G-space. For simplicity we may assume that $\mathrm{vert}(K) = [n]$ and vertex $i \in [n]$ is realized by $v_i \in \mathbb{R}^d$. Then the induced action is given by

$$G \times |K| \longrightarrow |K|,$$

$$\Big(g, \sum_{j \in J} t_j v_j\Big) \longmapsto \sum_{j \in J} t_j v_{g \circ j}.$$

Definition D.9. A G-action on a simplicial complex K is called *regular* if the following property is satisfied for each subgroup H of G:

(∗)   For each $g_0, \dots, g_n \in H$ and $\{v_0, \dots, v_n\} \in K$ such that $\{g_0 v_0, \dots, g_n v_n\} \in K$, there exists an element $g \in H$ such that $g v_i = g_i v_i$ for all $i = 0, \dots, n$.

A simplicial complex with a regular G-action is called a *regular G-complex*. Let L be a subcomplex of K. If G acts regularly on K and leaves L invariant, i.e., $g\sigma \in L$ for all $g \in G$ and $\sigma \in L$, then we call the pair (K, L) a *regular G-pair*.

Lemma D.10. *If K is a G-complex, then the induced action on its second barycentric subdivision* $\mathrm{sd}^2 K$ *is regular.*

Proof. The proof is the content of Exercises 12–15.   □

Quotient and Fixed-Point Spaces of Regular Actions

Now let K be a regular G-complex. We define the *quotient* K/G to be the simplicial complex on the vertex set $\{v^* : v \in \text{vert}(K)\}$, where the vertices v^* are defined to be the orbits $v^* = Gv = \{gv : g \in G\}$ of the vertices of K. The simplices of K/G are given by all sets $\{v_0^*, \ldots, v_n^*\}$, where $\{v_0, \ldots, v_n\} \in K$. Note that due to the regularity of the G-action, two simplices $\{v_0^*, \ldots, v_n^*\}$ and $\{w_0^*, \ldots, w_n^*\}$ are identical if and only if there exists a $g \in G$ such that $w_i = gv_i$ for all $i = 0, \ldots, n$. Hence the simplices of K/G are in one-to-one correspondence with the orbits of simplices of the action of G on K.

Lemma D.11. *If K is a regular G-complex, then there are homeomorphisms* $|K|/G \cong |K/G|$ *and* $|K^G| \cong |K|^G$.   □

Note that K^G may be considered a subcomplex of K/G via

$$\text{vert}(K^G) \longrightarrow \text{vert}(K/G),$$

$$v \longmapsto v^* = \{v\}.$$

$\mathbb{Z}_2$-Actions on Spheres

Lemma D.12. *Let a $\mathbb{Z}_2$-action on the n-sphere be given by a continuous map $v : \mathbb{S}^n \to \mathbb{S}^n$. If the action is free, i.e., v is fixed-point-free, then there exists a continuous $f : \mathbb{S}^n \to \mathbb{S}^n$ such that $f(v(x)) = -f(x)$.*

Proof. The map f defined by

$$f : \mathbb{S}^n \longrightarrow \mathbb{S}^n,$$

$$x \longmapsto \frac{x - v(x)}{\|x - v(x)\|}$$

clearly satisfies all requirements.   □

G-Equivariant Maps and Simplicial Approximation

Proposition D.13. *Let G be a finite group, and let X and Y be G-spaces. Assume that the action on X is free and that X possesses a G-invariant triangulation. If $\dim(X) \leq n$ and Y is at least $(n-1)$-connected, then there exists a G-equivariant map $f : X \to Y$.*

Proof. Without loss of generality we may assume that $X = |K|$ is the polyhedron of a free G-complex K. Denote the k-dimensional skeleton by $K^{(k)} = \{\tau :$ $\tau \in K, \dim(\tau) \leq k\}$. We construct the map f inductively on the k-skeletons. Let $f^{(-1)} = \emptyset$ be the empty map. Assume that a G-equivariant map $f^{(k-1)} :$ $|K^{(k-1)}| \to Y$ has already been defined for $n \geq k \geq 0$. Let $G\sigma_1, \ldots, G\sigma_l$ be the distinct G-orbits of the k-dimensional simplices of K. Then for each j, the map $f^{(k-1)}$ is already defined on the sphere $|\partial\sigma_j|$. Since $k-1 \leq n-1 \leq \mathrm{conn}(Y)$, there exists an extension $f^{(k)}|_{|\sigma_j|} : |\sigma_j| \to Y$. Now define $f^{(k)}|_{G|\sigma_j|}$ by the G-action, i.e., $f^{(k)}(gx) = gf^{(k)}(x)$ for all $g \in G$ and $x \in |\sigma_j|$. Then we have found an extension $f^{(k)} : |K^{(k)}| \to Y$.   □

Theorem D.14. *Let K be a G-complex, L a regular G-complex, and $f : |K| \to |L|$ a G-equivariant map. Then there exist an $r \geq 0$ and a G-equivariant simplicial approximation $g : \mathrm{sd}^r K \to L$ such that $|g| : |K| \to |L|$ is G-equivariantly homotopic to f.*   □

The regularity condition on L can slightly be weakened. We refer to [Bre72, p. 68].

Exercises

1. Show that if p is prime, then the nonzero elements of $\mathbb{Z}_p$ form a group under multiplication.
2. Show that if G is a finite group of even order, then it contains an element of order 2.
3. Let G be an abelian group of order mn with greatest common divisor $(m, n)=1$. Assume that there exist elements a and b of order m and n, respectively. Show that G is cyclic.
4. Let G be a cyclic group of order n, and k a divisor of n. Show that G has precisely one subgroup of order k.
5. Let G and H be finite cyclic groups. Show that the product $G \times H$ with componentwise multiplication is cyclic if and only if the greatest common divisor of their orders is $(|G|, |H|) = 1$.
6. (Euler–Fermat) Let a be an integer and p a prime not dividing a. Show that $a^{p-1} \equiv 1 \mod p$.
7. Let $H < G$ be a subgroup and $a \in G$. Show that aHa^{-1} is a subgroup of G that is isomorphic to H.
8. Let $H \trianglelefteq G$ be a normal subgroup. Show that the multiplication in the quotient group G/H as defined on page 210 is well defined and satisfies the group axioms.
9. Let $f : G \to H$ be a group homomorphism. Show that the preimage $f^{-1}(N)$ of a normal subgroup $N \trianglelefteq H$ is normal in G.

10. Let $f : G \to H$ be a homomorphism of groups. Assume that H is abelian and that $N < G$ is a subgroup containing ker f. Show that N is normal in G.

11. Show that a group action may be identified with a group homomorphism $G \to$ Sym(X) from G to the symmetric group of X.

12. Show that the following two properties of a G-action on a simplicial complex K are equivalent.

   (A) For each $g \in G$ and $\sigma \in K$, g leaves $\sigma \cap g\sigma$ pointwise fixed.
   (A') If v and gv belong to the same simplex, then $gv = v$.

13. Show that if K is a G-complex, then the induced action on $K' = $ sd K satisfies property (A) as defined in the previous exercise.

14. Show that if K is a G-complex satisfying property (A) as defined in Exercise 12, then the induced action on $K' = $ sd K satisfies the following property (B):

   (B) For each $g_0, \ldots, g_n \in G$ and $\{v_0, \ldots, v_n\} \in K$ such that $\{g_0 v_0, \ldots, g_n v_n\} \in K$, there exists a $g \in G$ such that $gv_i = g_i v_i$ for all $i = 0, \ldots, n$.

15. Prove Lemma D.10 employing the previous exercises.

16. Provide a proof of Lemma D.11.

17. Show that the map f in Lemma D.12 must be surjective. Give an example of an action v such that f is not injective.

Chapter 9
Appendix E: Some Results and Applications from Smith Theory

In this appendix we introduce Smith theory. It deals with group actions on simplicial complexes, their fixed points and orbit spaces, as well as what can be said about these in terms of homology theory.

The chapter is based on Glen E. Bredon's excellent book [Bre72] except for the final result, which is based on Robert Oliver's article [Oli75].

E.1   The Transfer Homomorphism

Let G be a finite group and K a regular G-complex. The oriented simplicial chain complex $C_*(K)$ inherits a G-action via $g\langle v_0, \dots, v_n \rangle = \langle gv_0, \dots, gv_n \rangle$. Similarly, if (K, L) is a regular G-pair, then $C_*(K, L)$ inherits a G-action. The group ring $\mathbb{Z}G$ consisting of formal sums $\sum_{g \in G} n_g g$, where $n_g \in \mathbb{Z}$, acts on $C_*(K)$, resp. $C_*(K, L)$, by $(\sum_{g \in G} n_g g)c = \sum_{g \in G} n_g (gc)$. The element $\sigma = \sum_{g \in G} g \in \mathbb{Z}G$ is called the *norm*. The norm σ yields a chain map $\sigma : C_*(K) \to C_*(K)$, and hence the image $\sigma C_*(K)$ is a subchain complex of $C_*(K)$. There is an analogous construction for $\sigma : C_*(K, L) \to C_*(K, L)$.

Now the canonical simplicial quotient map

$$K \longrightarrow K/G,$$

$$\{v_0, \dots, v_n\} \longrightarrow \{v_0^*, \dots, v_n^*\},$$

induces a homomorphism $\pi : C_*(K, L) \to C_*(K/G, L/G)$.

Lemma E.1. *The kernels of the homomorphisms σ and π coincide, i.e.,* $\ker \sigma = \ker \pi$.

Proof. Let $s_1, \dots, s_m$ be the orbit of a simplex s and let s^* denote the simplex of K/G that is the image of s under the canonical simplicial quotient map. Clearly it suffices to consider a chain of type $c = \sum_{i=1}^m n_i s_i$. If G_s denotes the stabilizer

M. de Longueville, *A Course in Topological Combinatorics*, Universitext,
DOI 10.1007/978-1-4419-7910-0_9,
© Springer Science+Business Media New York 2013

of s, i.e.,

$$G_s = \{g \in G : gs = s\},$$

then $m = |G|/|G_s|$ by Lemma D.7. Now

$$\pi(c) = \sum_{i=1}^{m} n_i \pi(s_i) = \sum_{i=1}^{m} n_i s^*,$$

and hence $\pi(c) = 0$ if and only if $\sum_{i=1}^{m} n_i = 0$. We now turn our attention to $\sigma(c)$. Since $g\sigma(c) = \sigma(c)$ for any $g \in G$, and since G acts transitively on the set of s_i, we obtain that $\sigma(c) = r \sum_{i=1}^{m} s_i$ for some integer r. Since the sums of coefficients on both sides of this equation have to be identical, we obtain

$$|G| \sum_{i=1}^{m} n_i = rm = r|G|/|G_s|.$$

Therefore $r = |G_s| \sum_{i=1}^{m} n_i$, and hence $\sigma(c) = 0$ if and only if $r = 0$ if and only if $\sum_{i=1}^{m} n_i = 0$. $\square$

Hence we obtain the following sequence:

$$C_*(K/G, L/G) \cong C_*(K, L)/\ker \pi = C_*(K, L)/\ker \sigma \cong \sigma C_*(K, L),$$

which yields a homomorphism

$$\mu : C_*(K/G, L/G) \longrightarrow C_*(K, L),$$

$$\pi(c) \longmapsto \sigma(c).$$

The induced homomorphism in homology

$$\mu_* : H_*(K/G, L/G) \longrightarrow H_*(K, L)$$

is called the *transfer*. Now, the canonical quotient map yields a homomorphism in the other direction,

$$\pi_* : H_*(K, L) \longrightarrow H_*(K/G, L/G).$$

For the compositions we clearly obtain

$$\pi_* \mu_* = |G| : H_*(K/G, L/G) \longrightarrow H_*(K/G, L/G),$$

$$\mu_* \pi_* = \sigma_* = \sum_{g \in G} g_* : H_*(K, L) \longrightarrow H_*(K, L).$$

Now for the image of μ, it is obvious that $\operatorname{im}\mu \subseteq C_*(K,L)^G$, and hence $\operatorname{im}\mu_* \subseteq H_*(K,L)^G$. Let us consider the composition $\mu_*\pi_*$ restricted to $H_*(K,L)^G$:

$$H_*(K,L)^G \xrightarrow{\pi_*} H_*(K/G, L/G) \xrightarrow{\mu_*} H_*(K,L)^G.$$

If $c \in C_*(K,L)^G$ is a cycle, then

$$\mu_*\pi_*([c]) = \sum_{g\in G} g_*[c] = [\sum_{g\in G} gc] = |G|[c].$$

In summary we obtain the following beautiful result.

Theorem E.2. *If F is a field of characteristic 0 or characteristic relatively prime to $|G|$, then*

$$\pi_* : H_*(K,L;F)^G \longrightarrow H_*(K/G, L/G;F)$$

and

$$\mu_* : H_*(K/G, L/G;F) \longrightarrow H_*(K,L;F)^G$$

are isomorphisms.   □

E.2 Transformations of Prime Order

We are now interested in the special situation in which the group G has prime order p and homology is computed with coefficients in the field $\mathbb{Z}_p$ of prime order p. Note that this is in stark contrast to the conditions in Theorem E.2. Let us consider the group G multiplicatively generated by the element g, i.e., $G = \langle g \rangle$. We will be interested in the elements $\tau = 1 - g$ and the norm $\sigma = 1 + g + g^2 + \cdots + g^{p-1}$ of the group ring $\mathbb{Z}_p G$. The significance of τ should be clear, since $c \in C_*(K,L)$ is an element of $C_*(K,L)^G$ if and only if $\tau c = 0$. Note that $\tau\sigma = 0 = \sigma\tau$ and that $\sigma = \tau^{p-1}$, since

$$(-1)^i \binom{p-1}{i} \equiv 1 \pmod{p}.$$

More generally, for $1 \le i \le p-1$ we want to consider $\rho = \tau^i$ and put $\bar{\rho} = \tau^{p-i}$. Then $\tau = \bar{\sigma}, \sigma = \bar{\tau}$, and $\rho\bar{\rho} = 0$.

Proposition E.3. *For each $\rho = \tau^i$, $1 \le i \le p-1$, the sequence of chain complexes*

$$0 \to \bar{\rho}C_*(K,L;\mathbb{Z}_p) \oplus C_*(K^G, L^G;\mathbb{Z}_p) \xrightarrow{i} C_*(K,L;\mathbb{Z}_p) \xrightarrow{\rho} \rho C_*(K,L;\mathbb{Z}_p) \to 0$$

is exact, where i denotes the sum of the inclusions and ρ : $C_*(K, L; \mathbb{Z}_p) \to$ $\rho C_*(K, L; \mathbb{Z}_p)$ is given by multiplication by ρ.

Proof. Consider the sequence in a fixed dimension n. It is easy to see that the sequence can be decomposed into two sequences

$$0 \to C_n(K^G, L^G; \mathbb{Z}_p) \xrightarrow{i} C_n(K^G, L^G; \mathbb{Z}_p) \xrightarrow{\rho} 0 \to 0$$

and

$$0 \to \bar{\rho} C_n(K, L; \mathbb{Z}_p) \xrightarrow{i} C \xrightarrow{\rho} \rho C_n(K, L; \mathbb{Z}_p) \to 0,$$

where $C \leq C_n(K, L; \mathbb{Z}_p)$ is the subgroup generated by all n-simplices $s \in K \setminus K^G$, $s \notin L$. For the exactness of the first sequence it suffices to note that for an n-simplex $s \in K^G$ we have $\tau(s) = 0$ and hence $\rho(s) = 0$.

Regarding the second sequence, it suffices to consider chains c consisting of formal sums of simplices that are contained in the orbit of a single simplex s. Since $s \notin K^G$, c may be written as $c = \sum_{i=0}^{p-1} n_i g^i s$, where $n_i \in \mathbb{Z}_p$. We can identify any such c with $\sum_{i=0}^{p-1} n_i g^i \in \mathbb{Z}_p G$, and hence the sequence that we are dealing with boils down to

$$0 \longrightarrow \bar{\rho}\Lambda \xrightarrow{i} \Lambda \xrightarrow{\rho} \rho\Lambda \longrightarrow 0,$$

where we denote the group ring $\mathbb{Z}_p G$ by Λ. Since $\rho \circ i = 0$ it suffices to show that $\dim \ker \rho = \dim \operatorname{im} i$ as $\mathbb{Z}_p$-vector spaces. Now $\dim \operatorname{im} i = \dim \bar{\rho}\Lambda$, $\dim \ker \rho = \dim \Lambda - \dim \operatorname{im} \rho = \dim \Lambda - \dim \rho\Lambda$, and hence we need to show that $\dim \rho\Lambda + \dim \bar{\rho}\Lambda = \dim \Lambda = p$. We are going to show that $\dim \tau^k \Lambda = \dim \Lambda - k = p - k$ for any $1 \leq k \leq p - 1$.

The kernel of $\tau : \Lambda \to \Lambda$ consists of formal sums with constant coefficients, i.e., multiples of $\sigma = 1 + \cdots + g^{p-1}$, and hence is 1-dimensional. But this kernel, $\mathbb{Z}_p \sigma$, is also contained in the image of any τ^j, $1 \leq j \leq p-1$, since $\sigma = \tau^{p-1} = \tau^j \tau^{p-1-j}$. Hence we are done by induction. $\square$

Definition E.4. Define the graded group $H_*^\rho(K, L; \mathbb{Z}_p) = H_*(\rho C_*(K, L; \mathbb{Z}_p))$ to be the *Smith special homology group*.

With this definition the long exact sequence that we obtain from Proposition E.3 yields the following result.

Proposition E.5. *For $\rho = \tau^i$ there is a long exact sequence in homology*

$$\xrightarrow{\partial} H_k^{\bar{\rho}}(K, L; \mathbb{Z}_p) \oplus H_k(K^G, L^G; \mathbb{Z}_p) \xrightarrow{i_*} H_k(K, L; \mathbb{Z}_p) \xrightarrow{\rho_*} H_k^\rho(K, L; \mathbb{Z}_p) \xrightarrow{\partial}$$

that we will refer to as the Smith sequence. $\square$

Note that an argument similar to the proof of Proposition E.3 yields the short exactness of the following sequence:

$$0 \to \sigma C_*(K, L; \mathbb{Z}_p) \xrightarrow{i} \tau^j C_*(K, L; \mathbb{Z}_p) \xrightarrow{\tau} \tau^{j+1} C_*(K, L; \mathbb{Z}_p) \to 0,$$

where $1 \leq j \leq p - 1$ and i denotes inclusion. Hence we obtain a long exact sequence.

Proposition E.6. *For $1 \leq j \leq p - 1$ there is a long exact sequence*

$$\xrightarrow{\partial} H_k^{\sigma}(K, L; \mathbb{Z}_p) \xrightarrow{i_*} H_k^{\tau^j}(K, L; \mathbb{Z}_p) \xrightarrow{\tau_*} H_k^{\tau^{j+1}}(K, L; \mathbb{Z}_p) \xrightarrow{\partial} . \qquad \square$$

As in the previous section, we want to understand the kernel of the homomorphism $\sigma : C_*(K, L; \mathbb{Z}_p) \to C_*(K, L; \mathbb{Z}_p)$. It should not differ too much from the case with integer coefficients except that $\sigma(s) = ps = 0$ whenever $s \in K^G$. More formally, the kernel of σ coincides with the kernel of the composition

$$C_*(K, L; \mathbb{Z}_p) \xrightarrow{j} C_*(K, K^G \cup L; \mathbb{Z}_p) \xrightarrow{\pi} C_*(K/G, K^G \cup L/G; \mathbb{Z}_p).$$

We leave the proof, which is very similar to the case of integer coefficients, to the exercises. We therefore obtain an isomorphism

$$\sigma C_*(K, L; \mathbb{Z}_p) \cong C_*(K/G, K^G \cup L/G; \mathbb{Z}_p)$$

inducing an isomorphism in homology.

Proposition E.7. *There is an isomorphism*

$$H_*^{\sigma}(K, L; \mathbb{Z}_p) \cong H_*(K/G, K^G \cup L/G; \mathbb{Z}_p). \qquad \square$$

E.3   A Dimension Estimate and the Euler Characteristic

Now we want to apply our previous results to obtain relations among the dimensions of the homology groups and the Euler characteristics of the considered spaces and chain complexes, respectively. As before, we will be concerned with group actions of prime order p and homology with coefficients in $\mathbb{Z}_p$.

Theorem E.8. *Let G be a cyclic group of prime order p, K a regular G-complex, and $L \subseteq K$ an invariant subcomplex. Then for any $n \geq 0$ and any $\rho = \tau^i$, $1 \leq i \leq p - 1$, the following inequality holds:*

$$\dim H_n^{\rho}(K, L; \mathbb{Z}_p) + \sum_{k \geq n} \dim H_k(K^G, L^G; \mathbb{Z}_p) \leq \sum_{k \geq n} \dim H_k(K, L; \mathbb{Z}_p).$$

Proof. This follows almost immediately from the Smith sequence. Note that in this sequence we may interchange ρ and $\bar{\rho}$. We will make use of both sequences that are obtained in this way. Let us first consider the sequence

$$H_{k+1}^{\bar{\rho}}(K, L; \mathbb{Z}_p) \overset{\partial}{\longrightarrow} H_k^{\rho}(K, L; \mathbb{Z}_p) \oplus H_k(K^G, L^G; \mathbb{Z}_p) \overset{i_*}{\longrightarrow} H_k(K, L; \mathbb{Z}_p).$$

First of all,

$$\dim H_k^{\rho}(K, L; \mathbb{Z}_p) + \dim H_k(K^G, L^G; \mathbb{Z}_p) = \dim \operatorname{im} i_* + \dim \ker i_*.$$

Clearly $\dim \operatorname{im} i_* \leq \dim H_k(K, L; \mathbb{Z}_p)$, and moreover, $\ker i_* = \operatorname{im} \partial$, and hence $\dim \ker i_* = \dim \operatorname{im} \partial \leq \dim H_{k+1}^{\bar{\rho}}(K, L; \mathbb{Z}_p)$. Substituting, we obtain the general inequality

$$\dim H_k^{\rho}(K, L; \mathbb{Z}_p) + \dim H_k(K^G, L^G; \mathbb{Z}_p)$$
$$\leq \dim H_k(K, L; \mathbb{Z}_p) + \dim H_{k+1}^{\bar{\rho}}(K, L; \mathbb{Z}_p).$$

Let us introduce the abbreviations

$$a_k = \dim H_k(K^G, L^G; \mathbb{Z}_p), \qquad b_k = \dim H_k(K, L; \mathbb{Z}_p),$$
$$c_k = \dim H_k^{\rho}(K, L; \mathbb{Z}_p), \qquad \bar{c}_k = \dim H_k^{\bar{\rho}}(K, L; \mathbb{Z}_p).$$

With this notation we obtain the two inequalities by interchanging ρ and $\bar{\rho}$:

$$c_k - \bar{c}_{k+1} + a_k \leq b_k \quad \text{and} \quad \bar{c}_k - c_{k+1} + a_k \leq b_k.$$

Consider the first inequality for $k = n, n+2, n+4, \ldots$ and the second for $k = n+1, n+3, n+5, \ldots$. The sum of all these gives the desired inequality $c_n + \sum_{k \geq n} a_k \leq \sum_{k \geq n} b_k$. $\qquad \square$

Euler Characteristic

Theorem E.9. *Let G be a cyclic group of prime order p, K a regular G-complex, and $L \subseteq K$ an invariant subcomplex. Then the following equality holds:*

$$\chi(K, L) + (p-1)\chi(K^G, L^G) = p\chi(K/G, L/G).$$

This in particular implies the congruence

$$\chi(K, L) \equiv \chi(K^G, L^G) \pmod{p}.$$

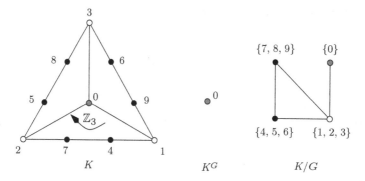

Fig. E.1 An illustration of Theorem E.9

An illustration of Theorem E.9 is given by Fig. E.1. In this case, $\mathbb{Z}_3$ acts regularly on the simplicial complex K by rotation about the center vertex by $120°$. The Euler characteristics of the respective complexes are $\chi(K) = 10 - 12 = -2$, $\chi(K^G) = 1$, and $\chi(K/G) = 4 - 4 = 0$. And indeed,

$$\chi(K) + (3 - 1)\chi(K^G) = 0 = 3\chi(K/G).$$

The following simple proof works for finite simplicial complexes. With some more effort it can be shown that the theorem holds more generally in the case that (K, L) and $H_*(K, L)$ are finite-dimensional; cf. [Bre72].

Proof. Recall that the Euler characteristic may be computed on the chain complex level, i.e.,

$$\chi(K, L) = \sum_{k \geq 0} (-1)^k \dim C_k(K, L).$$

We claim that the equation already holds in each dimension k. Consider an orbit $G\sigma$ of an n-simplex $\sigma \in K$, $\sigma \notin L$. It contributes 1 to the dimension of $C_k(K/G, L/G)$, and therefore $(-1)^k p$ to $p\chi(K/G, L/G)$.

Now either the orbit consists of p simplices, in which case $\sigma \notin K^G$, or it consists just of σ, in which case $\sigma \in K^G$. In the first case, the orbit of sigma contributes p to $\dim C_k(K, L)$ and 0 to $\dim C_k(K^G, L^G)$, and in the second case, it contributes 1 to $\dim C_k(K, L)$ and 1 to $\dim C_k(K^G, L^G)$. Hence the contribution to the left side of the equation $\chi(K, L) + (p - 1)\chi(K^G, L^G)$ is either $(-1)^k p + (p - 1)(-1)^k 0 = (-1)^k p$ or $(-1)^k \cdot 1 + (p - 1)(-1)^k \cdot 1 = (-1)^k p$. $\square$

E.4   Homology Spheres and Disks

Definition E.10. For $n \geq -1$, a simplicial complex K is called a *mod-p-homology n-sphere* if it has the same mod-p homology as the n-sphere, i.e., $H_*(K; \mathbb{Z}_p) \cong H_*(\mathbb{S}^n; \mathbb{Z}_p)$.

Theorem E.11. *If G is a p-group and K is a regular G-complex that is a mod-p-homology n-sphere, then the fixed-point complex K^G is a mod-p-homology r-sphere for some $-1 \leq r \leq n$. If p is odd, then $n - r$ is even.*

Note that the case in which the fixed-point complex is empty is covered, since the -1-sphere is the empty space.

Proof. Since the statement is trivial otherwise, we may assume that $n \geq 0$. We start with the special case in which G is a cyclic group of order p. The general case will then follow by induction. By Theorem E.8 we obtain

$$\sum_{k=0}^{\infty} \dim H_k(K^G; \mathbb{Z}_p) \leq \sum_{k=0}^{\infty} \dim H_k(K; \mathbb{Z}_p) = 2.$$

According to Theorem E.9, $\chi(K^G) \equiv \chi(K)$ (mod p). Hence it is impossible that $\sum_{k=0}^{\infty} \dim H_k(K^G; \mathbb{Z}_p) = 1$, since then $\chi(K^G) = \pm 1$, whereas $\chi(K)$ is either 0 or 2. Therefore $\sum_{k=0}^{\infty} \dim H_k(K^G; \mathbb{Z}_p)$ is either 0 or 2, proving that K^G is indeed a mod-p-homology r-sphere for some $r \geq -1$. The fact that $r \leq n$ is easy and relegated to the exercises. Clearly, if p is odd, the congruence $\chi(K^G) \equiv \chi(K)$ (mod p) implies that $n - r$ is even.

Now let's assume that G is a group of order p^k. We proceed by induction on k. Since G is a p-group, there exists a normal p-subgroup $G' \trianglelefteq G$ of order p^{k-1} such that G/G' is a cyclic group of order p, by Proposition D.8. By the induction hypothesis, $K^{G'}$ is a mod-p-homology r-sphere for some $-1 \leq r \leq n$, and if p is odd, then $n - r$ is even. Now consider the action of G/G' on $K^{G'}$. As we showed before, $K^G = (K^{G'})^{G/G'}$ is a mod-p-homology r'-sphere for some $-1 \leq r' \leq r \leq n$. Moreover, if p is odd, then $r - r'$ is even, and hence also $n - r' = (n-r) + (r-r')$ is even. $\qquad\square$

Definition E.12. A pair (K, L) of simplicial complexes is called a *mod-p-homology n-ball* if $H_k(K, L; \mathbb{Z}_p) = 0$ for $k \neq n$, and $H_n(K, L; \mathbb{Z}_p) \cong \mathbb{Z}_p$.

In other words, (K, L) is a mod-p-homology n-ball if it has the same mod-p homology as the pair $(\mathbb{B}^n, \mathbb{S}^n)$.

Theorem E.13. *If G is a p-group and (K, L) is a regular G-pair that is a mod-p-homology n-ball, then (K^G, L^G) is a mod-p-homology r-ball for some $0 \leq r \leq n$. If p is odd, then $n - r$ is even.*

Proof. The proof proceeds along the lines of the proof of Theorem E.11, and the details are left to the exercises. $\qquad\square$

For the following corollary, recall that a complex K is called mod-p acyclic if all reduced homology groups with $\mathbb{Z}_p$-coefficients vanish.

Corollary E.14. *If G is a p-group and K is a regular mod-p-acyclic G-complex, then the fixed-point complex K^G is mod-p acyclic as well.*

Proof. Set $L = \emptyset$ and $n = 0$. $\qquad\qquad\qquad\qquad\qquad\qquad\qquad\qquad\qquad$ $\square$

E.5 Cyclic Actions and a Result by Oliver

Lemma E.15. *Let G be a cyclic group of order n and K a regular G-complex. If K is $\mathbb{Q}$-acyclic, then $\chi(K^G) = 1$.*

Proof. We proceed by induction on n. Assume that $n = mp^k$ for some prime factor $p \mid n$, $k \geq 1$ and $p \nmid m$, and that the lemma has been proven for cyclic groups of order m. Let G be multiplicatively generated by the element g. We consider the subgroups $H = \langle g^{p^k} \rangle$ of order m, $H' = \langle g^m \rangle$ of order p^k, and $H'' = \langle g^{mp} \rangle$ of order p^{k-1}. Now set $L = K^H/H''$. Then the cyclic group H'/H'' of order p acts on L via $[h]H''x = H''(hx)$, which is well defined since any two representatives of an element of H'/H'' differ only by an element of H''. The fixed-point complex of this action turns out to be

$$L^{H'/H''} = (K^H/H'')^{H'/H''} = K^G,$$

and the quotient $L/(H'/H'') = (K^H/H'')/(H'/H'')$ can be identified with K^H/H' via

$$(H'/H'')H''v \longmapsto H'v.$$

The Euler characteristic formula from Theorem E.9 yields

$$\chi(L) + (p-1)\chi(L^{H'/H''}) = p\chi(L/(H'/H'')).$$

In other words,

$$\chi(K^G) = \frac{1}{p-1}\left(p\chi(K^H/H') - \chi(K^H/H'')\right).$$

Note that since $p \nmid m$, we have $K^H/H' = (K/H')^H$ and $K^H/H'' = (K/H'')^H$. In order to use the induction hypothesis for the action of H, we have to check that K/H' and K/H'' are $\mathbb{Q}$-acyclic. But this follows directly from the transfer isomorphism

$$\mu_* : H_*(K/\Gamma;\mathbb{Q}) \xrightarrow{\cong} H_*(K;\mathbb{Q})^\Gamma$$

for $\Gamma = H'$, respectively $\Gamma = H''$. So by the induction hypothesis, $\chi((K/H')^H) = 1$ and $\chi((K/H'')^H) = 1$. It follows that

$$\chi(K^G) = \frac{1}{p-1}(p-1) = 1. \qquad \qquad \square$$

We are finally able to prove the main result.

Theorem E.16. *Let G be a finite group with a normal subgroup $H \lhd G$ of order p^k for some prime p and $k \geq 1$, so that G/H is a cyclic group. If K is a mod-p-acyclic complex with regular G-action, then for the fixed-point complex, $\chi(K^G)=1$.*

Proof. By Corollary E.14, K^H is mod-p acyclic. By the *universal coefficient theorem* (see [Bre93]),

$$0 \to H_k(K^H;\mathbb{Z}) \otimes F \to H_k(K^H;F) \to H_{k-1}(K^H;\mathbb{Z}) * F \to 0$$

for $F = \mathbb{Z}_p$, we obtain that the homology $H_*(K^H;\mathbb{Z})$ has no free summands except in dimension 0. Then, for $F = \mathbb{Q}$, we obtain that K^H is also $\mathbb{Q}$-acyclic. Hence, by Lemma E.15, we obtain that $\chi(K^G) = \chi((K^H)^{G/H}) = 1$. $\qquad \square$

Exercises

1. Show that for $0 \leq i \leq p - 1$,

$$(-1)^i \binom{p-1}{i} \equiv 1 \pmod{p}.$$

2. Show that the kernel of the homomorphism $\sigma : C_*(K, L; \mathbb{Z}_p) \to C_*(K, L; \mathbb{Z}_p)$ coincides with the kernel of the composition

$$C_*(K, L; \mathbb{Z}_p) \xrightarrow{j} C_*(K, K^G \cup L; \mathbb{Z}_p) \xrightarrow{\pi} C_*(K/G, K^G \cup L/G; \mathbb{Z}_p).$$

3. In the proof of Theorem E.11, provide the details for the fact that $r \leq n$.
4. Provide a proof of Theorem E.13.

References

[Aig88]   Martin Aigner, *Combinatorial search*, Wiley-Teubner, 1988.

[Alo87]   Noga Alon, *Splitting necklaces*, Adv. Math. **63** (1987), 247–253.

[Arc95]   Dan Archdeacon, *The thrackle conjecture*, http://www.emba.uvm.edu/~archdeac/problems/thrackle.htm. Accessed August 2012.

[Arn49]   Bradford H. Arnold, *A topological proof of the fundamental theorem of algebra*, Am. Math. Monthly **56** (1949), 465–466.

[AS92]   Noga Alon and Joel H. Spencer, *The probabilistic method*, Wiley-Interscience Series in Discrete Mathematics and Optimization, John Wiley & Sons, Inc., New York, 1992.

[Bár78]   Imre Bárány, *A short proof of Kneser's conjecture*, J. Combinatorial Theory, Ser. A **25** (1978), 325–326.

[Bár79]   ———, *On a common generalization of Borsuk's and Radon's theorem*, Acta Math. Acad. Sci. Hung. **34** (1979), 347–350.

[BEBL74]   M. R. Best, P. van Emde Boas, and H. W. jun. Lenstra, *A sharpened version of the Aanderaa–Rosenberg conjecture*, Afd. zuivere Wisk. ZW **300/74** (1974).

[Bie92]   Thomas Bier, *A remark on Alexander duality and the disjunct join*, unpublished preprint, 7 pages, 1992.

[Bin64]   R. H. Bing, *Some aspects of the topology of 3-manifolds related to the Poincaré conjecture*, Lectures on Modern Mathematics (T.L. Saaty, ed.), vol. II, Wiley, 1964, pp. 93–128.

[Bir59]   Bryan John Birch, *On 3N points in a plane*, Math. Proc. Cambridge Phil. Soc. **55** (1959), 289–293.

[Bjö80]   Anders Björner, *Shellable and Cohen–Macaulay partially ordered sets*, Trans. AMS **260** (1980), no. 1, 159–183.

[Bjö94]   ———, *Topological methods*, Handbook of Combinatorics (R. Graham, M. Grötschel, and L. Lovász, eds.), North Holland, Amsterdam, 1994, pp. 1819–1872.

[BK07]   Eric Babson and Dmitry Kozlov, *Proof of the Lovász conjecture*, Annals of Mathematics (Second Series) **165** (2007), no. 3, 965–1007.

[Bol04]   Bela Bollobas, *Extremal graph theory*, ch. 8. Complexity and Packing, Dover Publications, Mineola, New York (2004).

[Bre72]   Glen E. Bredon, *Introduction to compact transformation groups*, Academic Press, New York-London (1972).

[Bre93]   ———, *Topology and geometry*, Graduate Texts in Mathematics, vol. 139, Springer-Verlag, 1993.

[Bre06]   Felix Breuer, *Gauss codes and thrackles*, Master's thesis, Freie Universität Berlin, 2006.

M. de Longueville, *A Course in Topological Combinatorics*, Universitext,
DOI 10.1007/978-1-4419-7910-0,
© Springer Science+Business Media New York 2013

[Bro03]   Torsten Bronger, *PP3 – Celestial Chart Generation*, http://pp3.sourceforge.net/, 2003. Accessed March 2010.

[BSS81]   Imre Bárány, Senya B. Schlosman, and András Szűcs, *On a topological generalization of a theorem of Tverberg*, J. London Math. Soc. **23** (1981), no. 2, 158–164.

[BW83]   Anders Björner and Michelle Wachs, *On lexicographically shellable posets*, Trans. AMS **277** (1983), no. 1, 323–341.

[CLSW04]   Péter Csorba, Carsten Lange, Ingo Schurr, and Arnold Wassmer, *Box complexes, neighborhood complexes, and the chromatic number*, J. Comb. Theory, Ser. A **108** (2004), no. 1, 159–168.

[CN00]   Grant Cairns and Yury Nikolayevsky, *Bounds for generalized thrackles*, Discrete Comput. Geom. **23** (2000), no. 2, 191–206.

[Die06]   Reinhard Diestel, *Graph theory*, 3rd ed., Graduate Texts in Mathematics, vol. 173, Springer, New York, 2006.

[Dol83]   Albrecht Dold, *Simple proofs of some Borsuk–Ulam results*, Contemp. Math. **19** (1983), 65–69.

[Fan52]   Ky Fan, *A generalization of Tucker's combinatorial lemma with topological applications*, Ann. Math. **56** (1952), no. 2, 431–437.

[FT81]   Robert M. Freund and Michael J. Todd, *A constructive proof of Tucker's combinatorial lemma*, J. Comb. Theory, Ser. A **30** (1981), 321–325.

[Gal56]   David Gale, *Neighboring vertices on a convex polyhedron*, Linear Inequalities and Related Systems (H. W. Kuhn and A. W. Tucker, eds.), Annals of Math. Studies, vol. 38, Princeton University Press, 1956, pp. 255–263.

[GR92]   J. E. Green and Richard D. Ringeisen, *Combinatorial drawings and thrackle surfaces*, Graph Theory, Combinatorics, and Algorithms **1** (1992), no. 2, 999–1009.

[Gre02]   Joshua E. Greene, *A new short proof of Kneser's conjecture*, Amer. Math. Monthly **109** (2002), 918–920.

[HT74]   John Hopcroft and Robert Tarjan, *Efficient planarity testing*, J. of the Assoc. for Computing Machinery **21** (1974), no. 4, 549–568.

[Hun74]   Thomas W. Hungerford, *Algebra*, Graduate Texts in Mathematics, vol. 73, Springer, 1974.

[JA01]   Elizabeth J. Jewell and Frank R. Abate (eds.), *New Oxford American Dictionary*, Oxford University Press, 2001.

[Jän84]   Klaus Jänich, *Topology*, Springer-Verlag New York, 1984.

[KK80]   D.J. Kleitman and D.J. Kwiatkowski, *Further results on the Aanderaa-Rosenberg conjecture*, J. Comb. Theory, Ser. B **28** (1980), 85–95.

[Kne55]   Martin Kneser, *Aufgabe 360*, Jahresbericht der Deutschen Mathematiker-Vereinigung **58** (1955), no. 2, 27.

[Koz07]   Dmitry N. Kozlov, *Chromatic numbers, morphism complexes, and Stiefel-Whitney characteristic classes*, Geometric combinatorics (Ezra Miller et al., ed.), 13, Institute for Advanced Studies. IAS/Park City Mathematics, American Mathematical Society; Princeton, NJ, 2007, pp. 249–315.

[KSS84]   Jeff Kahn, Michael Saks, and Dean Sturtevant, *A topological approach to evasiveness*, Combinatorica **4** (1984), no. 4, 297–306.

[Lon04]   Mark de Longueville, *Bier spheres and barycentric subdivision*, J. Comb. Theory, Ser. A **105** (2004), 355–357.

[Lov78]   László Lovász, *Kneser's conjecture, chromatic number and homotopy*, J. Combinatorial Theory, Ser. A **25** (1978), 319–324.

[LPS97]   László Lovász, János Pach, and Mario Szegedy, *On Conway's thrackle conjecture*, Discrete Comput. Geom. **18** (1997), no. 4, 369–376.

[Mas77]   William S. Massey, *Algebraic topology: An introduction*, Graduate Texts in Mathematics, vol. 56, Springer-Verlag, 1977.

[Mas91]   ———, *A basic course in algebraic topology*, Graduate Texts in Mathematics, vol. 127, Springer Verlag, 1991.

[Mat04] Jiří Matoušek, *A combinatorical proof of Kneser's conjecture*, Combinatorica **24** (2004), no. 1, 163–170.

[May99] J. Peter May, *A concise course in algebraic topology*, The University of Chicago Press, 1999.

[Meu05] Frédéric Meunier, *A $\mathbb{Z}_q$-Fan formula*, preprint (2005), 14 pages.

[Mun84] James R. Munkres, *Elements of algebraic topology*, Addison-Wesley, Menlo Park, California, 1984.

[MW76] Eric Charles Milner and Dominic J. A. Welsh, *On the computational complexity of graph theoretical properties*, Proc. 5th Br. comb. Conf. (Aberdeen 1975), 1976, pp. 471–487.

[MZ04] Jiří Matoušek and Günter M. Ziegler, *Topological lower bounds for the chromatic number: a hierarchy.*, Jahresbericht der Deutschen Mathematiker-Vereinigung **106** (2004), no. 2, 71–90.

[Oli75] Robert Oliver, *Fixed-point sets of group actions on finite acyclic complexes*, Comment. Math. Helv. 50 (1975), no. 50, 155–177.

[Ore67] Oystein Ore, *The four color problem*, Academic Press, New York, 1967.

[Öza87] Murad Özaydin, *Equivariant maps for the symmetric group*, unpublished preprint (1987), 17 pages, University of Wisconsin, Madison.

[Pak08] Igor Pak, *The discrete square peg problem*, http://arxiv.org/abs/0804.0657, 2008. Accessed June 2010

[Pól56] George Pólya, *On picture-writing*, Am. Math. Mon. **63** (1956), 689–697.

[PS05] Timothy Prescott and Francis Edward Su, *A constructive proof of Ky Fan's generalization of Tucker's lemma*, J. Comb. Theory, Ser. A **111** (2005), no. 2, 257–265.

[PSŠ07] Michael J. Pelsmajer, Marcus Schaefer, and Daniel Štefankovič, *Removing even crossings*, J. Comb. Theory, Ser. B **97** (2007), 489–500.

[Rot87] Gian-Carlo Rota, *The Lost Café*, Los Alamos Science, Special Issue (1987), 23–32.

[RV76] Ronald L. Rivest and Jean Vuillemin, *On recognizing graph properties from adjacency matrices*, Theor. Comput. Sci. **3** (1976), 371–384.

[RW98] Jack Robertson and William Webb, *Cake-cutting algorithms. Be fair if you can.*, A. K. Peters., 1998.

[Saa72] Thomas L. Saaty, *Thirteen colorful variations on Guthrie's Four–Color conjecture*, Amer. Math. Monthly **79** (1972), no. 1, 2–43.

[Sar91] Karanbir S. Sarkaria, *A one-dimensional Whitney trick and Kuratowski's graph planarity criterion*, Israel J. Math. **73** (1991), no. 1, 79–89.

[Sar92] ———, *Tverberg's theorem via number fields*, Israel J. Math. **79** (1992), no. 2-3, 317–320.

[Sar00] ———, *Tverberg partitions and Borsuk–Ulam theorems*, Pacific J. Math. **196** (2000), 231–241.

[Sch78] Alexander Schrijver, *Vertex critical subgraphs of Kneser graphs*, Nieuw Arch. Wiskd. **26** (1978), no. III., 454–461.

[Schu06] Carsten Schultz, *Graph colorings, spaces of edges and spaces of circuits*, Advances in Math. **221** (2006), no. 6, 1733–1756.

[Schu10] ———, *On the $\mathbb{Z}_2$-homotopy equivalence of $|\mathcal{L}(G)|$ and $|\operatorname{Hom}(K_2, G)|$*, personal communication, 2010.

[Schy06] Daria Schymura, *Über die Fragekomplexität von Mengen- und Grapheneigenschaften*, Master's thesis, Freie Universität Berlin, 2006.

[See06] K. Seetharaman, *Pretzel production and quality control*, Bakery Products: Science and Technology (Y. H. Hui, Harold Corke, Ingrid De Leyn, Wai-Kit Nip, and Nanna A. Cross, eds.), Blackwell, first ed., 2006, pp. 519–526.

[Spe28] Emanuel Sperner, *Neuer Beweis für die Invarianz der Dimensionszahl und des Gebietes*, Abhandlungen Hamburg **6** (1928), 265–272.

[SS03] Francis Edward Su and Forest W. Simmons, *Consensus-halving via theorems of Borsuk–Ulam and Tucker*, Math. Social Sci. **45** (2003), 15–25.

[ST07]   Gábor Simonyi and Gábor Tardos, *Colorful subgraphs in Kneser-like graphs*, European J. Combin. **28** (2007), no. 8, 2188–2200.

[Su99]   Francis Edward Su, *Rental harmony: Sperner's lemma in fair division*, Am. Math. Monthly **10** (1999), 930–942.

[Tho92]   Carsten Thomassen, *The Jordan–Schoenflies theorem and the classification of surfaces*, Amer. Math. Monthly **99** (1992), 116–130.

[Tho94]   _____, *Embeddings and minors*, Handbook of Combinatorics (R. Graham, M. Grötschel, and L. Lovász, eds.), North Holland, Amsterdam, 1994, pp. 301–349.

[Tho98]   Robin Thomas, *An update on the Four–Color theorem*, Notices of the Amer. Math. Soc. **45** (1998), no. 7, 848–859.

[Tut70]   William Thomas Tutte, *Toward a theory of crossing numbers*, J. Comb. Theory **8** (1970), 45–53.

[Tve66]   Helge Tverberg, *A generalization of Radon's theorem*, J. London Math. Soc. **41** (1966), 123–128.

[vK32]   Egbert R. van Kampen, *Komplexe in euklidischen Räumen*, Abh. Math. Semin. Hamb. Univ. **9** (1932), 72–78, 152–153.

[Vol96]   Alexey Yu. Volovikov, *On a topological generalization of the Tverberg theorem*, Math. Notes **59** (1996), no. 3, 324–326.

[Wal83]   James W. Walker, *From graphs to ortholattices and equivariant maps*, J. Comb. Theory, Ser. B **35** (1983), 171–192.

[Wes05]   Douglas B. West, *Introduction to graph theory*, Prentice-Hall, 2005.

[Woo71]   Douglas R. Woodall, *Thrackles and deadlock*, Proc. Conf. Combinatorial Mathematics and Its Applications, Oxford 1969 (1971), 335–347.

[Woo80]   _____, *Dividing a cake fairly*, J. Math. Anal. Appl. **78** (1980), 233–247.

[Zee63]   E.C. Zeeman, *On the dunce hat*, Topology **2** (1963), 341–358.

Index

M. de Longueville, *A Course in Topological Combinatorics*, Universitext,
DOI 10.1007/978-1-4419-7910-0,
© Springer Science+Business Media New York 2013